Carolina Rocks!

Carolina Rocks!
The Geology of South Carolina

Carolyn Hanna Murphy

Sandlapper Publishing Co., Inc.

Published by Sandlapper Publishing Co., Inc.
Orangeburg, South Carolina

Designed and Typeset by: Tiger Creek Productions,
Columbia, South Carolina
Typestyle: Times New Roman & Helvetica

Photographs, unless otherwise stated, are by the author.

Manufactured in the United States of America

Library of Congress Cataloging-in-Publication Data

Murphy, Carolyn Hanna, 1946–
    Carolina Rocks! : the geology of South Carolina / by Carolyn Hanna
Murphy.
        p.   cm.
    Includes bibliographical references (p. 231) and index.
    ISBN 0-87844-121-2
    1. Geology—South Carolina—Juvenile literature.   [1. Geology—
South Carolina. 2. South Carolina—Geography.]   I. Title.
QE161.M87   1995
557.57—dc20

                                                        94–20979
                                                            CIP

DEDICATION

To my parents, Helen and Bill Hanna
who gave me rocks, books, love, and encouragement

and

to my students at Hopkins and Summit Parkway
Middle Schools,
who have been the very best teachers . . .

CONTENTS

LIST OF TABLES

LIST OF PHOTOGRAPHS

ACKNOWLEDGMENTS

This book is the result of my personal quest as a teacher to better understand the geology of South Carolina and to make that information available to my students and to the interested public. *Carolina Rocks!* could not have been written without the help of many generous people who gave selflessly of their time, advice, materials, and criticism. Numerous people expressed the need for a book about South Carolina geology for the student and general reader including Dr. John Carpenter, director of the University of South Carolina's Center for Science Education and Dr. Phil Astwood and Dr. Donald Colquhoun, professors in the University of South Carolina's Department of Geology. I am especially grateful to Frank Handal of Sandlapper Publishing for his early belief in this book; to Barbara Stone, my editor; and Tiger Creek Productions, my copyeditors, who deserve special thanks for their patience, skill, and good advice throughout each stage of this project.

Many people from academic and government institutions helped to create this book by providing photographs, and offering revisions and advice. They include Daniel Fairey and Richard Lacy of the Land Resources and Conservation Division; Steve Bennett of the Non-Game and Heritage Trust Division of the South Carolina Department of Natural Resources; Ann Rose of the Natural Resources Soil Conservation Service; Dr. Nancy

Soderberg and Dr. Bill Dillon of the United States Geological Survey; Ralph Willoughby and Alan-Jon Zupan of the South Carolina Geological Survey; Dr. Joyce Bagwell of the Seismic Monitoring Network and Earthquake Education Center; Dr. Larry Brown of Cornell University; Dr. Robert Hatcher of the University of Tennessee; and Dr. Pradeep Talwani of the University of South Carolina.

I am grateful to the South Carolina State Museum for permission to use photographs from its extensive fossil collection. Natural History Curators Michael Ray and Jim Knight helped me with photographic production, fossil information, and continuous encouragement.

Jim and Carolyn Smoak of the South Carolina Earth Antiquities Society allowed me to photograph their extensive marine fossils and the Columbia Gem and Mineral Society permitted me to photograph specimens from their annual rock and mineral show. Geology hobbyists Frank Hill and Read Miner shared their petrified wood and mineral collections with me and provided extensive field trip information.

Coastal geologists Dr. Miles Hayes of Research Planning, Inc., and Dr. Tim Kana of Coastal Science and Engineering, Inc., helped me with photographs and advice concerning the barrier islands, and were more than generous with their time and materials.

The staffs of the University of South Carolina Thomas Cooper Map Library and the South Caroliniana Library were most helpful and timely in aiding my quest for books, photographs, and maps.

Several mining companies gave me tours of their property, allowed me to take pictures, and provided detailed information. Thanks to Andy Fisher of the Tarmac Granite Quarry in Columbia, Steve Hamilton of the Martin Marietta Granite Quarry in Cayce, John Hammer of U.S. Silica in Lexington County, W. H. Kirkland of the Huber Corporation in Langley, Patrick Highsmith of the Ridgeway Gold Mine in Fairfield County, and John Scheetz of the Brewer Mine in Chesterfield County.

I am thankful to Dr. Ed Hayes for his photographic assistance and encouragement, and for joining me in field trip research into muddy riverbeds and caves and across fields and woods when he would rather have been watching basketball.

I owe a large debt to my sister and mentor, Dr. Helen Black, who blazed her own trail through the world of education and invited me to follow; and lastly, I owe my daughter, Jennifer, special thanks and love for her unflagging belief in this project. She kept me on task—she did the dishes while I wrote, and she also fed the cats.

While many people helped me to produce this book by providing corrections, information, advice, graphics, and photographs, all errors, both of omission and of commission, are my own.

INTRODUCTION

Why study South Carolina geology?

Every school year this plaintive question is asked of middle school science teachers by puzzled students, and every year teachers search for a convincing response. The students continue in the following vein:

> Why should we study South Carolina geology when this state doesn't have any volcanoes, or big mountains like the Rockies, or a Grand Canyon? Compared to these places, South Carolina geology is *boring!*

Granted, South Carolina is not Colorado, with its snow-capped mountains that soar over 16,000 feet; it does not have active volcanoes like Hawaii; and it is not Arizona that does contain the grandest of canyons. But, South Carolina has its own interesting geology. Its geology is simply less showy, more subtle, than those famous and well-trampled western places! South Carolina geology just asks us to hunt a bit harder for our finds, that's all!

Actually, South Carolina possesses geological treasures that extend from one end of the state to the other. Gold and gemstones fit for a queen's crown are found here, ancient and colorful mountains that are older and were once taller than the Alps, ancient windblown dunes that contain the purest of silica sands, Carolina bays filled with rare wildlife and dark mysteries, remains of ancient volcanoes that once darkened the sky with ash, and lonely fossils that are still waiting to be found by students and taken home. So, while South Carolina is not Colorado or Arizona or Hawaii, it does contain enough interesting geology to teach and delight students for a lifetime.

If geology students are willing to hunt a bit and go outside their houses and school doors to pick up the sand, rocks, and soils that are found there, then the amazing story of South Carolina's geology will be within their reach. By traveling a bit farther, just a few miles down the road, they can experience the even greater excitement of seeing South Carolina both as it is today and as it was thousands and millions of years ago. Rocks, minerals, and fossils lay waiting to reveal their age-old stories.

South Carolina's libraries contain books, magazines, and maps that tell its geological story. South Carolina has museums with helpful experts who do historic field research, and there are fossil and mineral clubs that sponsor lectures and field trips open to all. Our universities and colleges offer courses on every aspect of geology. Many quarry and mine owners are generous and eager to share knowledge, and often provide access to their rocks and minerals for exploration, collection, and study.

This book is written to help you begin or continue your study of the geology of South Carolina. We'll explore how and when our state was made, what it is made of, and even where it might be going in the future! We'll look at some interesting historical events that happened in South Carolina that helped to shape our state, including the Carolina gold rush and the Charleston earthquake. Field trip sites that are especially interesting and accessible are noted for each section of the state. Information on government agencies, museums, rock clubs, rock shops, books, and individuals is found in the resources section at the end of the book. The bibliography lists books and articles about geology in general, and South Carolina geology in particular, which will be helpful to you and easy to find.

I hope that as you discover, or rediscover, the geology of South Carolina, you will enjoy and seek to preserve every aspect of this stunning and irreplaceable natural beauty—including the plants and animals that you encounter along the way. No hobby can be more delightful in this spectacular setting! I am confident that you will quickly see just how *un-boring* South Carolina's geology really is!

So, enjoy your exploration of the Palmetto State—the Grand Canyon and the Rocky Mountains can wait!

A Word About the Geologic Time Scale

Geologists and paleontologists use the geologic time scale as a reference in studying geological history, rocks, and fossils. The scale was developed during the nineteenth and twentieth centuries, and has been added to and revised over time as more information about the major geological events of the earth and the development of life have become known. Radiometric and other dating techniques helped in the development of the time scale because they made accurate dating of rocks and fossils possible, and these dating techniques continue to be refined.

The geologic time scale is divided into four eons that were named according to the general character of life that existed in each time period. The first eon, known as the Pre-Archean, dates from the formation of the earth—4.6 billion years ago—to 3.8 billion years ago when life first appeared in ancient rocks. The Archean eon (or Archeozoic, meaning *ancient life*) is also known as the Early Precambrian. It is the least understood of the earth's time periods due to its scarcity of fossils and the destruction of earth's earliest rocks. Dating from 3.8 billion to 2.5 billion years ago, this eon includes the development of microcontinents and the beginnings of primitive lifeforms. The third eon, the Proterozoic (meaning *former life*), began 2.5 billion years ago and lasted until about 570 million years ago. It included that time when large continents formed and soft-bodied animals first developed.

The fourth eon, the Phanerozoic (meaning *visible life*), dates from 570 million years ago to the present. It is divided into three eras, which describe life as ancient (Paleozoic), middle (Mesozoic), and recent (Cenozoic).

The three eras of the Phanerozoic are divided into periods that represent expanses of time divided by major disturbances in the earth's crust, such as mountain building. These periods are named either for the location where formations of its age were first studied or are well exposed or for a particular characteristic of those formations. For instance, the first three periods (the Cambrian, Ordovician, and the Silurian) are named after Wales (Cambria) and two of its ancient tribes. The Devonian period was named for the county of Devon in England. Wales and England were sites of important European geological work during the last century. The Mississippian and Pennsylvanian periods were named for two states in America, and the Permian period for a region of Russia. The Cretaceous period was named after the white chalk cliffs of southeastern England—which include the White Cliffs of Dover. These unique names given to the Paleozoic periods of the geologic time scale are reminders of the early development of geology because, in addition to marking changes in tectonics and biology, they also bear witness to the early pioneers in the fields of geology and paleontology.

The periods of the geologic time scale are divided into epochs which represent distinct times based upon fossil correlations. Finally, the epochs are divided into stages or ages. For further clarification, each unit on the scale—eon, era, period, epoch, or age—can be broken down into even smaller phases: early, middle, and late. Because the expanses of time are too great to know the exact years when major geological events occurred or when animals and plants appeared or disappeared, ranges of time must be used in dating the geological and paleontological past.

The geologic time scale is part of the scientific international language and is used by geologists all over the world. Although perhaps confusing at first, with practice and repeated reference to it, the geologic time scale will become helpful and convenient to use.

GEOLOGIC TIME SCALE

Millions of Years Before Present	Era	Period	Epoch	Tectonics	Major Life Forms
0.01	CENOZOIC	Quaternary	Holocene	S.C.: Charleston Earthquake; Barrier Islands form	Rise of human civilizations; extinction of some large mammals, such as mastodons
1.6			Pleistocene	Sea levels rise and fall; Ice Ages; S.C.: Carolina Bays form Climate is colder and drier	S.C.: Sabre-toothed tigers, mastodons, wolves, camels, horses, mammoths; great grazing herds of antelope & bison roam the state; Native Americans arrive from Asia
5.3		Tertiary	Pliocene	Cascadian Orogeny in the West Grand Canyon begins to form S.C.: Sea levels rise: Orangeburg Scarp forms; modern Alps & Himalayas rise	Great migration of animals: camels, horses, elephants, & bears migrate to South America from North America; sloths, armadillos, & glyptodonts come north.
22.7			Miocene	S.C.: Sandhills form; Gulf Stream develops off coast shaping continental shelf; Phosphates deposited	S.C.: Climate cools; many animal species become extinct.
36.8			Oligocene	S.C.: Seas retreat as Appalachians uplifted; erosion increases	Explosive diversification of life begins to slow; marine faunal families decrease.
57.8			Eocene	S.C.: Sea levels rise; deposits of marine sediments; Land bridges form	Whales diversify; early primates develop. S.C.: many marine fossils deposited on coastal plain such as sea turtles, alligators, & whales.
66.4			Paleocene	S.C.: Earthquakes become common; South Georgia Embayment and Cape Fear Arch begin to form	Rise of grasses, fruits & grains; Mammal diversification
144	MESOZOIC	Cretaceous		Asteroid hits Yucatan Beginning of the Rocky Mountains & the Andes; North America separates from Eurasia S.C.: Extensive clay beds deposited	Extinction of dinosaurs, ammonoids; development of flowering plants. S.C.: Hadrosaur dies near Kingstree, buried in river sediments.

Geologic Time Scale

Millions of Years Before Present	Era	Period	Epoch	Tectonics	Major Life Forms
208	MESOZOIC cont.	Jurassic		Orogeny in North America: volcanoes & mountain-building in the West; S.C.: North America rotates toward West and pulls away from Africa; Diabase dikes form, running N, NE, NW	First birds; dinosaurs dominant.
245		Triassic		S.C.: Rift valleys form; Crowburg Basin fills with sediments; Pangaea begins to break up	First primitive mammals; dinosaurs develop.
286	PALEOZOIC	Permian		S.C.: Blue Ridge shoved 160 miles northwest; extensive overthrusting; End of Alleghanian Orogeny	Permian Extinction: 50 percent of all species become extinct, including trilobites.
320		Pennsylvanian		Alleghenian Orogeny	Coal swamps; first reptiles; abundant insects.
360		Mississippian		Erosion	Amphibians dominant; seed plants develop
408		Devonian		Acadian Orogeny; S.C.: Metamorphic foliation of rocks	First amphibians; fish dominant.
438		Silurian		Extensive erosion	First land plants; corals dominant.
505		Ordovician		Taconic Orogeny: Inner Piedmont and Avalon terrain welded into North America	First fish; invertebrates dominant.
570		Cambrian		S.C. near the equator	Trilobites, brachiopods, & other marine invertebrates; Burgess shale fossils deposited; Batesburg trilobites deposited on island arc.
570 to 4.6 billion		Precambrian		Several mountain-building episodes; Iapetus Ocean forms 700 - 800 million years before present; Grenville Orogeny 1.2 billion years before present	Marine invertebrates develop; Bacteria, blue-green algae; micro fossils; oldest fossils found are 3.8 billion years old.

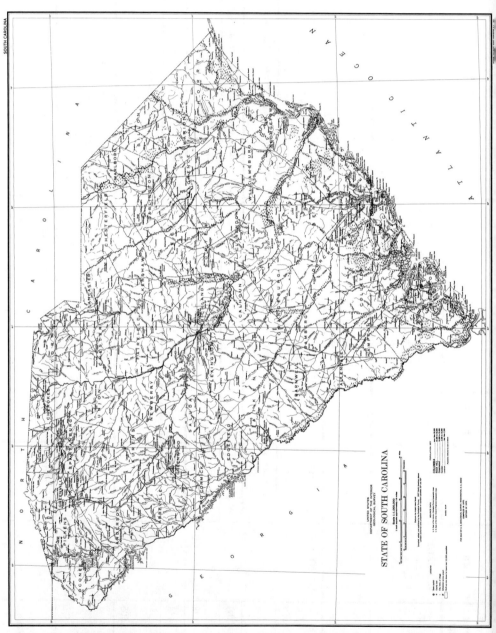

Map of the state of South Carolina (courtesy: United States Department of the Interior, Geological Survey)

CHAPTER 1

THE GEOGRAPHY OF SOUTH CAROLINA

South Carolina is one of America's smallest states. It ranks fortieth of the fifty states in size and is the smallest of the Deep South states. Although it covers only 19.5 million acres (31,113 square miles in area), the diversity of the landforms within this heart-shaped state is striking. South Carolina has been divided into five distinct geographic regions by land use. These regions, which closely parallel the geology of the state, lie in bands running southwest to northeast and include the Blue Ridge, the Piedmont, the Sandhills, the Coastal Plain, and the Coastal Zone. A traveler can drive north to south in the course of four or five hours and cross completely different terrains—from the rugged mountains of the Blue Ridge, over the rolling hills of the Piedmont, and up and over the beginnings of the Coastal Plain at the Sandhills. Moving south and southeast, one can view the rapids of the Fall Zone; the fields, forests, and swamps of the Coastal Plain; and then the salt marshes, and the barrier and sea islands of the Coastal Zone.

The present topography of South Carolina began to form millions of years ago by the collision of continents, the uplift of mountains, the transgressions and regressions of ancient seas, erosion, and deposition. These processes are still shaping South Carolina's landscape in the present day as sediment is eroded off the mountains and hills and transported across the Piedmont

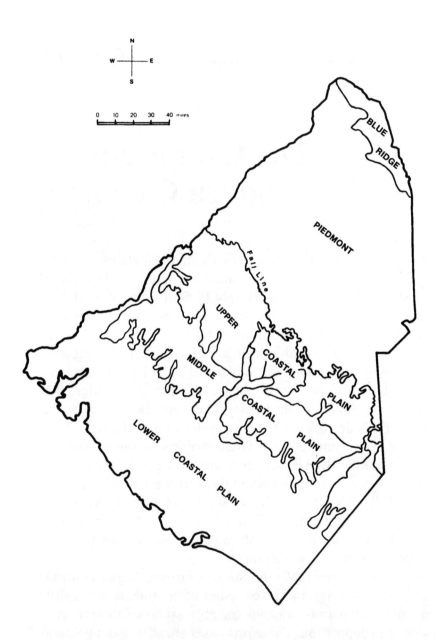

Landform Regions Map (courtesy: Facts on File)

and the Coastal Plain by rivers that flow to the Atlantic Ocean. South Carolina's temperate, humid climate averages 62 degrees Fahrenheit with 48 inches of rainfall annually. This moist climate accelerates the erosional processes that are continuously working to break down and carry away the mountains of the northwest and turn them into the soils and beach sands of the land to the south and southeast. Only the hardness of their rocks and the luxuriant growth of their forests protect the mountains from being more rapidly worn down to sea level.

Erosion of sediments on the upper Coastal Plain near Fort Jackson, Richland County

THE BLUE RIDGE REGION

The Blue Ridge Mountains run from northern Virginia to Georgia and cross the northwestern corner of South Carolina in Oconee and Pickens Counties. They are part of the larger Appalachian Mountain chain that stretches from Canada to Alabama. The Blue Ridge Mountains once stood thousands of feet higher, but today these rounded old ridges rise in South Carolina only a little higher than 3,400 feet.

With the highest annual rainfall in the state (60 inches per year), the Blue Ridge is the source of many mountain streams and rivers. Some of these streams and rivers give rise to spectacular waterfalls, such as Whitewater Falls, which begins in North Carolina and tumbles 400 feet into South Carolina. Although this region forms only 2 percent of the total land surface of South Carolina, geologically and geographically it is one of the most interesting regions to visit. Metamorphic outcrops are clearly exposed along road cuts and river banks, and mountain streams are swift and clear, cutting *V*-shaped valleys in the hard rock as they rush toward sea level. The rivers are shallow and filled with rocky shoals; a striking example is the Chatooga River on the Georgia/South Carolina border.

Chatooga River, South Carolina upcountry (courtesy: South Carolina Department of Parks, Recreation, and Tourism, Tourism Division)

While the Blue Ridge is a geologically distinct area bounded on the southeast by the Brevard Fault, geographically it includes both the ancient rocks of the Blue Ridge Mountains that lie in the northwestern corner of the state and the highlands of the Inner Piedmont belt that are not geologically related to the Blue Ridge at all. These highlands include hills and mountains that are found in Oconee, Pickens, and Greenville Counties such as Glassy Mountain, Table Rock, and Caesar's Head. Many mountains lie along the Cherokee Foothills Scenic Highway (S.C. Highway 11), north of Greenville. This area provides campers, visitors, and rock hounds with spectacular views and attracts multitudes of tourists. The highest peak in the state is Sassafras Mountain that straddles the North Carolina border and stands 3,554 feet above sea level. Pinnacle Mountain, with an elevation of 3,425 feet, is the highest peak that lies entirely within South Carolina.

Inner Piedmont highlands, South Carolina upcountry (courtesy: South Carolina Department of Parks, Recreation, and Tourism, Tourism Division)

THE PIEDMONT REGION

Below the Blue Ridge Mountains lies the Piedmont—a hundred-mile-wide expanse of dissected plain making up about one-third of the land area of South Carolina. The Piedmont begins at the twelve-hundred-foot level at the base of the mountains. Continuing toward the south and southeast, the Piedmont's elevation decreases and the topography becomes one of hills and valleys cut by streams that flow into the broad, muddy rivers that drain the Piedmont. These rivers include the Savannah, Broad, Saluda, and Catawba that flow southeast toward the Fall Zone from their headwaters in North Carolina. This great watershed (or drainage area) is formed by large quantities of rain and snow that fall in North Carolina as weather fronts that hit the Appalachian Mountains rise and drop moisture across the area. The runoff moves quickly because the soils of the Piedmont are mostly clays that overlie ancient basement rocks and are nearly impermeable. Once saturated, these clay soils repel rainwater that then moves downhill carrying sediments that fill the rivers with muddy water for much of the year.

Many geologists believe that the southern Piedmont was a peneplain in its distant past—a nearly featureless, gently undulating, and eroded plain that was later elevated and eroded again. The rivers in the Piedmont form dendritic stream patterns that cut into this upraised peneplain, moving upwards toward relatively narrow interstream divides, or high areas between the streams. Fertile floodplains are found along the large rivers, but most of the Piedmont is in slope—a highly eroded, mature terrain. The region is dotted with remnants of ancient mountains (or monadnocks) such as Little Mountain in Newberry County, Glassy Mountain in Pickens County, and Paris Mountain in Greenville County.

THE SANDHILLS

Much of the Piedmont is overlaid in South Carolina by the Sandhills. The Sandhills region—which, in some places, rises to 725 feet above sea level and up to 200 feet above the Piedmont—forms a geographic boundary between the Piedmont and the Coastal Plain. While geologically part of the Coastal Plain, the Sandhills are distinctive geographically. This region is made up of hilly, discontinuous bands of sand that trend northeast to southwest and are 30 miles wide in places. They run through portions of Chesterfield, Kershaw, Richland, Lexington, and Aiken Counties. Much of the sand of these Sandhills was blown into dunes during the Miocene, about 9 to 12 million years ago. Other sediments found in the Sandhills include ancient river deposits and weathered clays. Some of these sands and clays lie directly on top of the crystalline rocks of the Piedmont. The loose, permeable soils of the Sandhills do not hold rainwater well. As a result, areas with low water tables form a kind of semidesert in the middle of South Carolina. This can perhaps best be seen at the Sand Hill State Forest on U.S. Highway 1 near the town of Patrick in Chesterfield County. The plants that grow in these dry, relatively infertile soils are tenacious survivors that include turkey oak, longleaf pines, various cacti, briars, and berries. Homes and farms built in the Sandhills have gardens and fields that dry out quickly after a rainstorm. With irrigation and patient coaxing, however, fertile gardens, fine quality peaches, and abundant truck crops of vegetables can be produced in the Sandhills.

As the fast-moving rivers of the Piedmont meet the softer sediments of the Coastal Plain below the Sandhills, they make a sudden descent from an upland to a lowland area and form the rapids that mark the Fall Zone, which is over 1.5 miles wide in places. The cities of Columbia, North Augusta, Camden, and Cheraw lie within the Fall Zone. The eighteenth-century

The Fall Zone rapids on the Saluda River near the Riverbanks Zoo, Columbia

founders of these cities took advantage of the water power generated by the rapids to operate mills and establish industry. Because the rapids marked the most northerly points that could be reached by boat from the coast, profitable river trade developed. Later, river travel was replaced by the faster and more efficient railroads. Today, these Fall Zone cities are still centers of growing populations, trade, industry, and government. Farther up the eastern coast of the United States, the Fall Zone is home to some of America's most scenic and historic cities, including Raleigh, North Carolina; Richmond, Virginia; Washington, D.C.; and Philadelphia, Pennsylvania.

THE COASTAL PLAIN REGION

South of the Fall Zone, the topography changes dramatically with the broad expanse of the Coastal Plain. The largest of the geographic regions of the state, the Coastal Plain extends 120 to 150 miles from the Sandhills to the coast, covering about two-thirds of the state. Because the geography and the geology of this region allow for farming, quarrying, recreation, and convenient transportation, many of the important towns and cities of South Carolina are located here. They include: Charleston, Aiken, Georgetown, Beaufort, Orangeburg, Cheraw, Barnwell, Florence, Sumter, Marion, Manning, Myrtle Beach, and much of Columbia.

Much of the hilly upper Coastal Plain is forested with fast-growing pines and hardwoods, but large portions are used for farming. The upper Coastal Plain ends at ancient escarpments (or cliff faces) made by seas that receded long ago. The middle and lower Coastal Plain continues as an almost flat, featureless plain—interspersed with swamps—that stretches to the coast.

The great rivers of the Coastal Plain—such as the Congaree, Savannah, and Pee Dee—have vast floodplains and dark swamps along their edges. They meander extensively, forming oxbow lakes over time as they wend their way through huge quantities of sediments eroded from Piedmont rocks and deposited in the

Coastal Plain. Sandbars and spits are common along these rivers as the currents shift the sediment loads. The Great Pee Dee River, with headwaters in North Carolina, dominates the eastern part of the eastern Coastal Plain. It joins the Black River, the Little Pee Dee, the Lumber, the Pocotaligo, the Lynches, and the Waccamaw to drain much of the Coastal Plain into Winyah Bay. The Congaree and Wateree Rivers meet to form the Santee River that flows to the Atlantic Ocean between Georgetown and Charleston. The Savannah River drains the western region of South Carolina into the ocean near Hilton Head.

Savannah River meanders on the Coastal Plain, Allendale County. Note the oxbow lake at bottom. (courtesy: University of South Carolina, Thomas Cooper Map Library)

Human technology has created 525,000 acres of lakes in South Carolina that provide recreation for boating, camping, and fishing, and produce hydroelectric power for millions of people. The largest human-made lakes include Lake Hartwell, Lake Richard B. Russell, and Lake Thurmond (J. Strom Thurmond Lake) on the Savannah River; Lake Wylie on the Catawba River; Lake Wateree on the Wateree River; Lake Greenwood and Lake Murray on the Saluda River; and Lake Marion and Lake Moultrie on the Santee and Cooper Rivers. The Great Pee Dee River is the only one of the three major river systems that has been left undammed in South Carolina.

J. Strom Thurmond Lake, U.S. Highway 221 at Strom Thurmond Dam in Sumter National Forest

The Carolina bays, oval depressions that contain wetlands or savannahs, are unique to the middle and lower Coastal Plain. There are thousands of bays in South Carolina—big and small. Because their rich sediments are attractive for farming, most have been altered and exploited by agriculture. But today, some of these Carolina bays have become protected by parks and can now be explored and enjoyed by visitors.

THE COASTAL ZONE

The Coastal Zone is that part of the Coastal Plain extending about 10 miles inland from the coast. The landscape here changes dramatically as the savannahs, dark swamps, and coastal rivers that rise and fall with the ocean tides dominate the view. A wide band of marshes and oak trees greets travelers when thin arms of the sea reach inland as creeks that flow through salt marshes, hammocks, and swampy forests all along South Carolina's 183-mile coastline. As rivers meet the sea at the coast, they deposit rich silts in their estuaries that mingle with the saltier ocean water to provide an environment abundant in nutrients and capable of supporting large amounts of sea life and waterfowl. The delicate wetlands in the Coastal Zone are some of the most unspoiled wetlands on the eastern coast, and they are increasingly appreciated as one of South Carolina's greatest natural resources. One such area is the ACE Basin, so called because it is formed at the estuaries of a small river system made up of the Ashepoo, Combahee, and Edisto Rivers. Other rivers with headwaters in the Coastal Plain—the Black, Coosawhatchie, Combahee, and Salkehatchie—are black rivers that carry little sediment, but form rare ecosystems.

Coastal towns and cities built along these rivers, marshes, and creeks, and on the islands—including Charleston, Beaufort, Hilton Head, Georgetown, and Myrtle Beach—cater extensively to tourism.

Coastal Zone Regions

The coast is divided geographically into three zones. The first one, known as the Grand Strand, forms an arc of strand beaches that extends 60 miles from the North Carolina border south to Winyah Bay in Georgetown. This stretch is made up largely of eroding Pleistocene beach sediments with a few barrier islands, such as Pawleys Island, remaining along the coastline. The historic Waccamaw River runs parallel to the coast behind the Strand. About 250 years ago the river's protected shores were molded and dammed to form the rice plantations that provided South Carolina with much of its tremendous wealth and influence during the colonial era.

The second section of the coastline continues for 18 miles and includes Winyah Bay and the Santee Delta. Here the bays and islands form rare estuarine and marine habitats for hundreds of wildlife species. The Cape Romain National Wildlife Refuge, situated at the Delta, is one of the most pristine natural habitats remaining in the state.

The third section of the Coastal Zone is made up of the Sea Island Complex that begins at the Santee River Delta and runs for over 100 miles (continuing into Georgia). The sea islands were once part of the mainland but were cut off and isolated as sea levels rose during the Pleistocene. Some sea islands, such as Kiawah and Edisto, have barrier islands attached to them, while others, such as Hilton Head, have Holocene sand beaches. Some sea islands are bounded on their landward sides by deep embayments. St. Helena and Port Royal Sounds were once river valleys that were drowned by the rising sea. Unlike barrier island

Edisto Island marsh on the coast of South Carolina

Crabbing and shrimping on St. Helena Island (courtesy: South Carolina Department of Parks, Recreation, and Tourism, Tourism Division)

Heron in salt marsh (courtesy: South Carolina Department of Parks, Recreation, and Tourism, Tourism Division)

complexes farther north, South Carolina's sea islands face the Atlantic Ocean to the south and southeast and are therefore more sheltered from storms blowing in from the northeast, severe currents, and extreme high tides than the sea islands of neighboring states.

The Spanish moss-draped trees of the Coastal Zone announce to visitors that they have neared the limit of South Carolina's lands above water. Geologically the state continues miles out to sea under the continental shelf, but geographically it ends at the coast.

Seaoats atop dunes at Folly Island

In the following eight chapters, we will explore South Carolina by going back in time and place to peer under the earth's surface, bringing layers of buried rocks into view. In our imagination we can watch continents being pushed and pulled by forces deep within the earth. We can picture island arcs and lost microcontinents moving over the surface of the earth, then colliding with larger continents like drifted wood. We need to picture time not in years but in millions of years, and imagine places totally alien: land forested by no trees and a sea inhabited by few living organisms except algae, bacteria, protozoans, and soft-bodied invertebrates.

When South Carolina's geological history began, vertebrates—including people—did not exist. Their appearance had to await events millions of years in the future. Indeed, South Carolina began to form far from its present location over a half billion years ago . . .

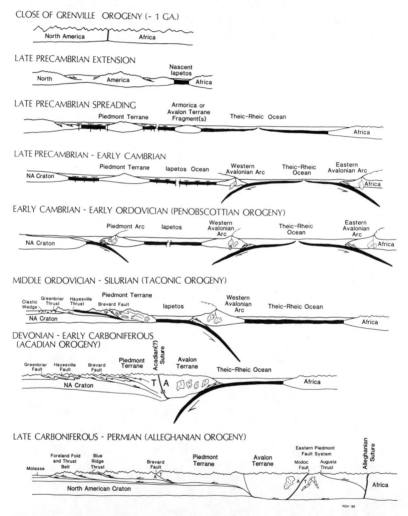

CLOSE OF GRENVILLE OROGENY (~ 1 GA.)

North America   Africa

LATE PRECAMBRIAN EXTENSION

Nascent
Iapetos
North   America   Africa

LATE PRECAMBRIAN SPREADING

Armorica or
Avalon Terrane
Fragment(s)
Piedmont Terrane   Theic–Rheic Ocean
                                                              Africa

LATE PRECAMBRIAN - EARLY CAMBRIAN

Piedmont Terrane   Iapetos Ocean   Western   Theic–Rheic   Eastern
                                                    Avalonian Arc   Ocean   Avalonian Arc
NA Craton   Africa

EARLY CAMBRIAN - EARLY ORDOVICIAN (PENOBSCOTTIAN OROGENY)

Piedmont Arc   Iapetos   Western   Theic–Rheic   Eastern
                                              Avalonian   Ocean   Avalonian
                                              Arc   Arc
NA Craton   Africa

MIDDLE ORDOVICIAN - SILURIAN (TACONIC OROGENY)

Piedmont Terrane
Clastic Greenbrier Hayesville   Western
Wedge   Thrust   Thrust   Avalonian
                   Brevard Fault   Iapetos   Arc   Theic–Rheic Ocean
NA Craton   Africa

DEVONIAN - EARLY CARBONIFEROUS
(ACADIAN OROGENY)

Greenbrier Hayesville Brevard   Piedmont   Avalon
Fault   Fault   Fault   Terrane   Terrane   Theic–Rheic Ocean
                                          T A   Africa
NA Craton

LATE CARBONIFEROUS - PERMIAN (ALLEGHANIAN OROGENY)

Eastern Piedmont
Fault System
Foreland Fold   Blue   Modoc   Augusta
and Thrust   Ridge   Brevard   Piedmont   Avalon   Fault   Thrust
Molasse   Belt   Thrust   Fault   Terrane   Terrane   Africa
North American Craton

ROH '86

Tectonics of the Southern and Central Appalachian Internodes, Robert
Hatcher (credit: reproduced with permission from the *Annual Review of
Earth and Planetary Sciences,* vol. 15, © 1987, by Annual Reviews, Inc.)

CHAPTER 2

FORMATION OF SOUTH CAROLINA

South Carolina's geographic structure closely reflects its geology. The state is divided into nearly parallel bands that extend from the northeast to the southwest. The Blue Ridge, in the northwest, is the highest part of the state. Continuing toward the coast are the Piedmont and the Coastal Plain. Each of these regions is geologically distinct. Each one tells a detailed story about the formation of South Carolina and how its geology fits into the overall puzzle of the formation and evolution of the earth's continents and oceans.

Although the mountains of the southeastern United States are not as tall and grand as the Rockies, these mountains have served as geological laboratories giving earth scientists a monumental puzzle to solve. The Appalachian and the Blue Ridge Mountains have provided important clues about the formation of mountain ranges and the development of continents. Significant questions have been raised and now largely answered: Why are the Appalachian Mountains wrinkled like a carpet pushed across a floor, with folds so distinct that they can be easily seen from space? Why are the ridges of the Blue Ridge Mountains stacked on top of the Piedmont like cordwood? How did the older metamorphic rocks of the Blue Ridge end up on top of younger igneous and sedimentary rocks that lie beneath them? And, what force could have produced such an effect? When geologists finally found answers to these questions they also confirmed the theory of mountain building, called plate tectonics, which is widely accepted today.

During the past 25 years, geologists have found evidence that South Carolina owes its existence, in part, to a continental fragment and a large island arc that collided with the North American coast beginning in the Ordovician Period. This collision raised a high mountain range that has since eroded. Then, in the late Paleozoic, the North American and African plates collided, which thrust the Blue Ridge and Piedmont rocks over each other, and pushed them against the previously flat-lying sediments of the Appalachian plateau. This collisional process folded the sediments into the Appalachian Mountains.

This tremendous accretion process heated and twisted the rocks until most of them became highly metamorphosed. The heat of subduction and collision created magma that flowed and

Landsat photo of the Appalachian Mountains. These folded mountains were uplifted as Africa and North America collided (330 to 270 million years ago). (courtesy: National Aeronautics and Space Administration)

then cooled far underground to form masses of granites; some of these granites, as well as sedimentary rocks, were further altered to become metamorphic rocks such as schists and gneisses. As the land rebounded over millions of years, these plutons and batholiths gradually rose toward the surface, and were exposed by erosion. Examples include Caesar's Head, Table Rock, and Forty Acre Rock.

The collision between North America and Africa was echoed throughout the world as all of the continents came together at the end of the Paleozoic to form the supercontinent of Pangaea. These continents stayed together for about 100 million years and then began to split apart during the Triassic. As they drifted apart, Europe and Africa left behind land that became permanently attached to North America. These pieces of African terrain make up most of Florida, parts of both southern Georgia and southern Alabama, and a large part of the land under the southeastern continental shelf. Parts of the North American plate became permanently attached to Europe and Africa as well, and today these North American fragments are found in such disparate sites as North Wales and Scotland in Great Britain, and Mauritania in West Africa.

After Africa and North America separated, South Carolina was left with a torn, faulted, and broken surface. The rifting, or breaking away, of these continents caused magma to well up due to the thinning of the rifting crust. Some of that magma took the form of narrow dikes (igneous rocks injected, while molten, into cracks in the older surrounding rocks). An example of one such dike is the large Flat Creek diabase dike in Lancaster County.

As North America and Africa drifted apart, fault-created valleys (called grabens) formed along the continents' edges. These grabens are now located far under the surface and are filled with hundreds of feet of sediments. As the once-stretched-and-heated land continued to adjust and rebound, it bent, sagged, and fractured. Many geologists feel that this process created a Seismic Zone near Charleston that even today is prone to earthquakes.

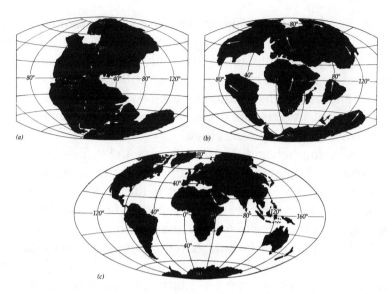

Map of the formation of Pangaea (courtesy: Dale E. Ingmanson and William J. Wallace, *Oceanography: An Introduction* [Belmont, Calif.: Wadsworth Publishing Co., 1989] 58)

About 550 million years ago the Palmetto State was still only a gleam in Mother Nature's eye. Although a smaller North American continent had existed for billions of years, it took pieces of continents and island arcs—combined with collisions by Africa, Baltica (early Northern Europe), and perhaps South America—to form South Carolina. Essentially, the continents were squeezed on a grand scale between a rock and a hard place! A great deal of evidence now exists to support a grand process of continent-building in which South Carolina played a part. Having begun as a hunch in the late nineteenth century, the idea has now developed into a theory that is largely accepted by scientists: plate tectonics.

PLATE TECTONICS

In the mid-1960s geologists and geophysicists discovered the "Rosetta stone" of geology. Just as the original Rosetta stone provided the key that enabled scholars to finally unlock the meaning of Egyptian hieroglyphics, the theory of plate tectonics provided the master idea that has enabled scientists to decipher hundreds of geological mysteries. An understanding of plate tectonics helps geologists explain major surface features such as continents, ocean basins, mountain ranges, volcanoes, and earthquakes. The theory also helps account for the occurrences of minerals and rocks, and the major processes that form the earth's surface and subsurface.

Plate tectonics is the conceptual framework upon which the geological history of South Carolina stands. As recently as 30 years ago, the growth of continents and the building of mountains was thought to be a sort of rebounding of sediments that had been eroded from the continents and deposited along the continental margins. Most geologists, at that time, thought that these heavy sediments pushed down the continental shelves and in the process formed a trough—called a geosyncline. It was believed that mountain chains were made from these sediments and associated volcanics, which rebounded along the continental edges over time. While sediments do gather in huge troughs along some continental shelves, and are indeed forming there today, the idea that mountains were built by the rebounding of these sediments was proved inaccurate with the development of plate tectonics theory in the 1960s. It became clear that mountains could not rise 25,000 feet or higher without more energy applied to them than crustal rebound. A tremendous lateral force would be needed to create mountain chains. The plate tectonics theory proposed that folded mountains are created when crustal plates collide with each other.

The plate tectonics theory suggests that the earth does not have one continuous, permanent, and inflexible crust, but rather that the earth's crust is made up of eight large plates and about a dozen smaller ones. These plates move at various rates around the earth—some almost continuously. The energy that drives these moving plates is thought to be provided by convection currents that rise and fall within the upper mantle—the hot, plastic layer 60 miles below the surface. Just as a convection current can be created in a sunny room when heated air rises (because it is lighter than the surrounding air) and the colder air sinks (because it is denser and heavier than the warmer air), so too can a convection current be created within the earth. The plastic rock material of the upper mantle rises as it warms and falls as it cools. This rising and falling of material within the mantle has a profound effect on the earth's surface. It is thought that as the convection currents converge and diverge far underground, they rift plates apart and move them along the surface. Then, as the top of the convection cells finally cools and falls, the convection currents drag heavy oceanic crust down into the mantle. Deep structures, called trenches, form at the edges of the continents where the oceanic crust is pulled downward. Heat increases with depth and the downward motion of the plates creates friction, so that at about 60 miles below the surface the rock melts and magma rises toward the surface forming volcanic lava. The rafting of these mobile plates over millions of years creates mountain ranges, island arcs, trenches, rift valleys, volcanoes, and earthquakes—giving rise to, or influencing, nearly all of the geological processes and features that occur on the earth.

The earth formed about 4.6 billion years ago, and, as it cooled, denser elements were pulled by gravity toward the center to form its large nickel-iron core. Lighter elements rose toward the surface to form a thin "scum" of continental material on the surface as early as 800 million years after the earth's formation. This differentiation of elements continues today as magma wells up to form the mid-ocean ridges at divergent plate boundaries and is subducted at the trenches at convergent plate boundaries. The heavier elements sink deeper into the earth and the lighter ones remain at, or near, the surface. These lighter elements then form igneous rocks (such as granites and andesites) and other lighter sedimentary and metaphoric continental rocks.

The source of the mantle's heat is largely the decay of radioactive elements such as uranium. As radioactive elements decay, they give off heat. If the earth were to last long enough to allow its store of these radioactive elements to be used in the decay process, the mantle would cool and the plate tectonics process would end.

When two continental plates collide, the continental crust wrinkles to form mountain ranges because the less dense continental crust cannot sink far downward within the denser rock of the oceanic crust. Light rock "floats" on dense rock in the same way that a cork floats on water. Because the cork is lighter than water, if it is pushed under the surface it will pop back up to the top. Continental crust reacts the same way. Since neither of the two colliding continents can go far down, both of them must go up! As the continents collide, their rocks crumple and form overthrusts and folds like a rug shoved across a floor.

Early evidence of plate movements included the similar coastlines between continents—an obvious example being the similarity between the western coastline of Africa and the eastern coastline of South America. Geographers and explorers as early as the sixteenth century recognized the peculiar fit of these two coasts and believed that the coastlines looked like pieces of a matching jigsaw puzzle. Further geological investigation and the development of plate tectonics theory provided evidence that Africa and South America not only had

Thrusted rock in the Great Smoky Mountains of North Carolina. Such thrusts moved South Carolina's rocks northwestward during the Appalachian Orogeny.

matching jigsaw coastlines, but that they also shared identical rock types, fossils, and mountain ranges in many areas along those coastlines—indicating that the continents were once joined.

Much of the evidence for plate movements was gathered in the late 1950s and the 1960s. Through worldwide exploration, experimentation, and the sharing of data between nations, scientists found that the vast majority of earthquakes and volcanoes occurred where plates collided. Deep sea magnetic readings showed that on either side of the mid-oceanic rift zone lava was poured out of undersea mountain ranges. Additionally, the oceanic crust ran parallel in magnetic direction and age away from the ridges. These readings also offered a startling discovery: although scientists knew that the earth was billions of years old, no oceanic crust could be found that was older than 200 million years. Where had it gone? It was discovered that oceanic crust was continually destroyed as it was subducted into the mantle at the edges of plates and that the ocean floors serve as conveyer belts for moving rocks.

Continental formations have occurred sporadically over time. The earth appears to leap into violent action, perhaps as heat builds up to a critical point in the asthenosphere, which then changes the flow of its convection currents. Great continent-forming events occurred 2.9 to 2.6 billion years ago, then at 1.9 to 1.7 billion years ago, and again at 1.1 to .9 billion years ago. The most recent continent-building episode began about 600 million years ago and resulted in the formation of mountain ranges like the Appalachians, Alps, Rockies, and Himalayas.

TECTONIC HISTORY OF SOUTH CAROLINA

The oldest rocks found in North America are located at its central core, or craton, found in central Canada and in the upper Midwest of the United States. Highly metamorphosed gneiss has been dated in Minnesota at 3.6 billion years and in Greenland at 3.8 billion years. Beginning about 1.2 billion years to 800

million years ago, an earlier continental collision with Africa
upraised the Grenville mountain range along North America's
eastern coast. It is at the base of this long-eroded, ancient mountain range that South Carolina began to form—in the northwestern corner of the state, at what is now the trailing edge of
the Blue Ridge thrust sheet. In Oconee County, and across the
Chatooga River into Georgia, lie Precambrian gneisses—
Grenville basement rocks that mark the former edge of the North
American continent before a continental fragment and exotic
terrains were attached to it during the Paleozoic.

Many geologists believe that North America had been joined
with Africa at least one other time (and perhaps several other
times) before the continents joined again 270 million years ago
to form the supercontinent, Pangaea. Then, beginning in the
Late Precambrian (about 800 million years ago), North America
split away from Africa. An ocean formed between the two continents that geologists call the Iapetus (in Greek mythology,
Iapetus was the father of Atlantis). It is at this rift zone that the
beginnings of the Blue Ridge Mountains are found. The range
began as a series of sediment-filled rift valleys, blocks of
Grenville crust pulled seaward, intruding volcanics, and, finally,
limestone that built up over the subsiding continental shelf.
These rocks and sediments waited millions of years for collision.

Then, in the Early Paleozoic about 560 million years ago, the
rifting ceased, reversed itself, and the Iapetus Ocean began to
close, bringing the continents together again. Neither the breaking apart nor the collision of the continents was a smooth movement. On a scale of millions of years this process was jerky,
bumpy, and intermittent. During most of the Paleozoic, though,
subduction, volcanism, folding, and thrust faulting were normal as mountains were raised and Pangaea assembled itself.
Three waves of orogeny, or mountain-building episodes, occurred along North America's eastern coast during the Paleozoic—the Taconic, Acadian, and Alleghenian.

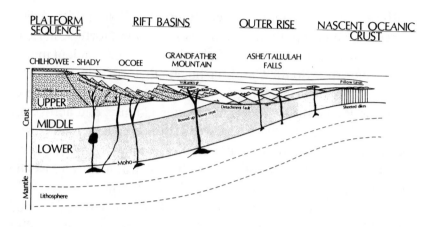

Grenville Rifting. Late Proterozoic/Early Paleozoic development of the eastern margin of North America applying the model of Lister and others. The Blue Ridge formed when the Iapetus Ocean closed. (courtesy: Hatcher and Goldberg, 1991)

The Taconic Orogeny

The Taconic Orogeny occurred in the Middle Paleozoic (470 to 440 million years ago) along the eastern coast of North America. It is thought that Northern Europe (then known as Baltica) and North America (known as Laurentia) collided, squeezing between them Precambrian exotic terrains that built up the mountains of Newfoundland, New England, and New York. Pieces of the largest of these exotic terrains (known as the Avalon terrain) have also been found across the Atlantic Ocean in Ireland, England, and Wales. Most geologists feel that North America also collided in the southeast with the Avalon terrain as the subduction between North America and the African plate powered a collision. Other geologists feel that North America collided with what is now the western region of South America. Research continues to determine how Africa and South America may have affected the southeastern coast of North America during the Ordovician. What is evident is that the

Avalon terrain, thought to be an island arc or even a microcontinent, was pressed against the North American plate by subduction. Its forward movement scraped up the diverse rocks and sediments of the Grenville coast that had intruded and been deposited on the continental shelf as Iapetus first opened in the Late Precambrian. These rocks became the Blue Ridge Mountains as they were shoved landward. Today in the Blue Ridge, these offshore sediments and rocks—metagreywackes, metasandstones, and metavolcanics, and their more highly metamorphosed equivalents of quartzites, schists, and gneisses—can be seen. For the first 25 million years, the Blue Ridge rose thousands of feet at the rapid pace of about 3.2 inches per century. Later, the pace slowed to about an inch per century.

The movement of the Avalon terrain toward North America is thought by some geologists to have also attached a large continental fragment that had split off from North America as the Iapetus opened. This reattached continental piece formed what is now known as the Inner Piedmont belt of South and North Carolina, Georgia, and Virginia. Inner Piedmont towns in South Carolina include Greenville, Spartanburg, Clemson, and Seneca. The Avalon terrain became fully attached to North America and formed most of the rest of the Piedmont—known as the Carolina terrain. This bedrock continues under the sediments of the Coastal Plain and the continental shelf.

Strong evidence linking Piedmont rocks to the Avalon terrain and Africa was discovered in 1982 when Middle Cambrian *Paradoxides* trilobite fossils were found in slightly metamorphosed mudstones, called metamudstones, near Batesburg in Lexington County. Trilobites were common marine arthropods that lived throughout the Paleozoic. The *Paradoxides* trilobites were not related to North American species, but they proved to be akin to trilobites found in Bohemia. Bohemia, although now attached to southern Europe, was once welded to North Africa. The presence of these very un-North American trilobites in South Carolina confirmed that the rock in which they were found was exotic and had been formed near Africa.

The Carolina terrain is divided into three belts: the Kings Mountain belt, the Charlotte belt, and the Carolina slate belt. Although each belt varies in its general metamorphic grade and in the frequency of its plutonic intrusions, all three belts are thought to be regions of a continuous piece of land. Towns such as Blacksburg, Laurens, and Lowndesville lie in the Kings Mountain belt; Rock Hill, Union, and York lie in the Charlotte belt; and, Saluda, Ridgeway, and Lancaster lie in the Carolina slate belt.

As the various terrains were attached to North America, the force of the collisions caused heating and high-grade regional metamorphism of many of the rocks. This led to the melting of rock far below the surface and caused the emplacement of granitic plutons miles underground. Tremendous thrust faulting formed high mountains, and then South Carolina, newly formed, became relatively quiet for about 100 million years.

The Acadian Orogeny

Pangaea continued to form around the world as continents moved toward each other. A repeated collision of Baltica and Laurentia, known as the Acadian Orogeny, followed 100 million years after the end of the Taconic Orogeny in the Late Devonian (380 to 340 million years ago). This collision affected the northern Appalachians more so than the southern Appalachians, but there is evidence of metamorphic foliation in some of the Piedmont rocks during this time. Subduction continued off the South Carolina coast as Africa and North America moved closer to each other.

The Alleghenian Orogeny

The final Paleozoic continent-building event—the Alleghenian Orogeny—occurred from the Pennsylvanian to Permian Periods (330 to 270 million years ago) as Africa and the other continents moved together to form the supercontinent Pangaea. At this time South Carolina was at the equator tipped about ninety degrees

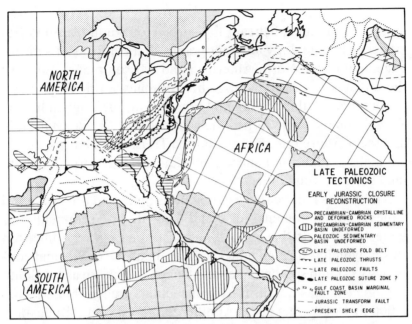

Diagram: Late Paleozoic tectonic features around the Atlantic Ocean (courtesy: Klitgard and Schouten, U.S. Geological Survey)

from its present position. The force that welded the continents was so powerful that the Blue Ridge and Piedmont were pushed northwestward, the Blue Ridge as far as 160 miles over the Piedmont rocks. This movement did not happen all at once, but was persistent over 80 million years. It thrust and buckled the Piedmont and the Blue Ridge toward the west. Geologists believe that the most intensive thrusting and regional metamorphism of this mountain-building episode took place in only 5 million years. By the end of this orogeny, Pangaea was complete.

THE AFRICA-NORTH AMERICA DISENGAGEMENT

Africa and North America remained together this time for about 70 million years. During the Late Triassic Period, North America began to pull away, eventually rotating westward to its present position.

In the sediments beneath the surface of South Carolina lies evidence of a geology and topography that confirms this disengagement of the North American and African plates. Because these plates were welded together for many millions of years, the break up of this continental bond created geological havoc. Today, this havoc underlies the sediments of the Coastal Plain and the offshore region in the buried Triassic rift basins, Jurassic diabase dikes and sheets, and faulted and warped ancient land surfaces found there.

Rift basins are created as crust pulls apart and down-faulted valleys, called grabens, form and then fill with sediment. The largest of the rift basins in South Carolina is the Dunbarton Basin, which is located under the Savannah River Site and continues west into southern Georgia where it is known as the South Georgia Rift Basin. This rift basin contains grabens buried from 6,000 to 14,000 feet deep. Another South Carolina rift basin is found farther to the east under Florence. Only the tiny Crowburg Basin near Pageland can be seen in South Carolina at the surface (see chapter 5).

The Late Triassic and Jurassic Periods witnessed a large outpouring of magma as the rifting of the continents continued for over 50 million years. At Charleston, Georgetown, Lake Moultrie, and west to Allendale, Coastal Plain sediments overlie a large diabase sheet called the "*J*-layer" (*J* for Jurassic), dated at about 180 million years. The *J*-layer lies on top of rift valleys and under hundreds of feet of Coastal Plain sediments. The magma that formed the *J*-layer radiated out in swarms and sheets from volcanoes that now lie buried offshore under the sediments of the continental shelf. Dikes radiated toward the north and northwest into the Piedmont and Blue Ridge. These dikes are dated at 195 to 205 million years, or Mid-Triassic. In Columbia, one such dike can be seen in Arcadia Lakes and a striking diabase dike is found on the west wall of the Martin Marietta Granite Quarry in Cayce. Because diabase is a mafic igneous rock that wells up as plates rift apart, these sheets and dikes give striking evidence of the violent movement that began

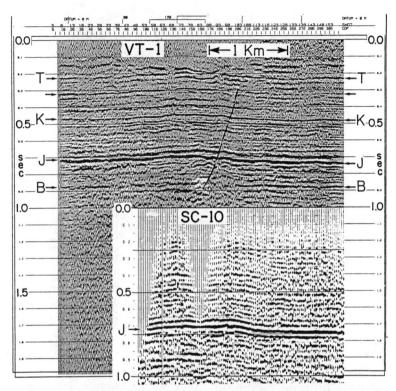

Rocks under Charleston showing *J*-layer. Virginia Tech VIBROSEIS line VT-1 in the Charleston area (B = basement; J = Basalt). (courtesy: Consortium for Continental Reflection Profiling [COCORP], Dr. Larry Brown)

Martin Marietta Granite Quarry in Cayce shows a diabase dike on the west wall

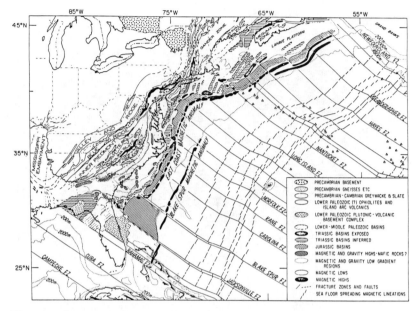

Diagram: Present Atlantic margin of eastern North America during the Paleozoic Era (courtesy: Klitgard and Schouten, U.S. Geological Survey, 1980)

in the Late Triassic and continued into the Jurassic Period, as
the supercontinent Pangaea pulled apart.

Steeply dipping Piedmont rocks lie under the sediments of
the Coastal Plain—sediments that are up to 4,500 feet in depth
beneath the western coast under Hilton Head. These Cretaceous
to Holocene sediments were deposited as a result of the erosion
of the ancient mountains of the Piedmont and Blue Ridge.

As Africa and North America slowly rifted apart over millions of years, the young, shallow Atlantic Ocean grew between
them. From the Mesozoic until late in the Cenozoic, as the sea
transgressed on the Coastal Plain, limestone formed. As the sea
regressed, clays and sands were deposited by rivers until a huge
wedge of sediments formed over the worn Piedmont rocks and
the filled rift valleys under the Coastal Plain. Isostatic rebound
to the north and northwest raised the level of the land. To the
south and southeast, the land cooled and sank. This cooling,
combined with the weight of the sediments themselves, caused
the entire Coastal Plain and the continental shelf to the south to

gradually sink. As the land offshore sank, limestone thousands of feet thick built up into what is now known as the Blake Plateau, 200 miles off the coast south of Charleston. During the past decade, the Blake Plateau has been explored by oil companies as a possible site of future hydrocarbon production. At the seaward end of the Blake Plateau is a huge escarpment called the Blake-Bahamas Scarp—the highest geological structure east of the Rockies—which descends 18,000 feet into the abyssal plain below. Many geologists believe that the boundary between the transitional African plate-stretched Piedmont crust and the thinner, but heavier, Jurassic oceanic crust lies under this ridge.

To the north of the South Georgia Rift Basin in both Georgia and South Carolina lay crystalline Piedmont rocks. However, southern Georgia and northern Florida are not made from Piedmont rocks, rather they were formed from pieces of exotic terrain—a microcontinent largely composed of African granite basement and Paleozoic felsic volcanic rocks. Both the rock layers and the fossils—the strata and the fauna—of this area can be correlated with those of the Bové Basin of Senegal in West Africa. This small piece of continent was attached to North America during the Alleghenian Orogeny and then it was left when the North American plate detached from the African plate (beginning in the Triassic). The suspected point of union (or suture) between Africa and North America is thought by many to run along the South Georgia Rift Basin—from southern Alabama through southern Georgia to southwestern South Carolina. Others believe the suture to be along the Brunswick

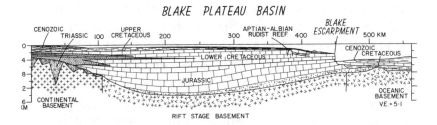

Idealized section across Blake Plateau Basin (courtesy: Dillon & Popenoe, U.S. Geological Survey, 1988)

Magnetic Anomaly. The anomaly, which runs west to east from Georgia offshore to the continental shelf south of the South Carolina coast, shows evidence of a vastly different rock type present at depth.

From the Late Cretaceous to the Miocene, the Gulf Trough Current (the precursor of the Gulf Stream), flowed from the Gulf of Mexico across southern Georgia to South Carolina. It eroded the sediments on the continental shelf off the coast and formed a trough. Then, as the current ended during the Miocene and the Gulf Stream began to flow around Florida, sediments from South Carolina flowed onto the continental shelf and began to build it up. This shelf is now one of the largest continental shelves in the world, extending out from the mainland well over 200 miles. Ocean currents have played a very important role in shaping the continental shelf and the coastline. Today, no carbonate banks are forming because the Gulf Stream effectively sweeps the shelf off the South Carolina coast. Eddies from the Gulf Stream continue to shape the scooped-out coastline along the Grand Strand and north to Cape Hatteras, North Carolina.

Geologists have estimated that the mountains formed by the Paleozoic collisions in North Carolina, Tennessee, Virginia, and South Carolina stood as high as 29,000 feet—surpassing the Swiss Alps and rivaling the Himalayas in height and grandeur. Subduction at its coasts has ended and South Carolina now lies on a stable, passive margin, trailing at the end of the North American continent. The erosion of these grand mountains has continued for millions of years and today their sediments form the soils of the Piedmont and the Coastal Plain, and the sands of the beaches and the continental shelf. No subduction zones are active along North America's eastern coast and no continents are headed its way from the east or south. North America has reached its present position by moving about a third of an inch or more per year, and it continues to move gradually toward the southwest. It is estimated that if nothing deflects its path, North America will hit east Asia in about 50 million years.

If new mountains are to be raised in future eons, they will probably first occur on North America's western edge.

Off South Carolina's coast is a wedge of sediments 2 to 3 miles thick. A new geosyncline is building—awaiting the day when another continental collision will result in orogeny. South Carolina lies ready to change course, as it has done over millions of years, propelled by the fickle convection currents of the mantle!

Table Rock, Pickens County

CHAPTER 3

GEOLOGY OF THE BLUE RIDGE AND PIEDMONT REGIONS

The Blue Ridge and the Piedmont provide South Carolina with large numbers of interesting and accessible rock formations and a wide variety of minerals. These beautiful regions are geologically very exciting, and once understood and appreciated, they tell a remarkable story of the subduction, mountain building, mineralization, erosion, and sedimentation that created the state. The geological history of the Blue Ridge and Piedmont has been largely deciphered using evidence provided by minerals, rocks, and large geological structures such as faults, anticlines, thrust sheets, and plutons.

MINERALS AND ROCKS OF THE BLUE RIDGE AND PIEDMONT REGIONS

The igneous and metamorphic rocks of the Piedmont and the Blue Ridge contain the greatest variety of minerals found anywhere in the state. This is due primarily to the complexity of the forces that formed them: collisions, thrusting, intrusions, folding, and recrystallization of original minerals during metamorphic episodes.

Over eighty minerals have been identified in South Carolina (see the gemstone list in appendix B). These minerals are found in the rocks of the Blue Ridge and Piedmont, and provide the rocks with a highly variable chemical composition. Some of the more common minerals include:

actinolite	galena	pyrophyllite
agate	garnet	pyroxene
allanite	goethite	pyrrhotite
almandine	gold	quartz (several
garnet	graphite	varieties)
alunite	hematite	riebeckite
andalusite	heulandite	rutile
apatite	hornblende	sapphire
aquamarine	ilmenite	serpentinite
arsenopyrite	jasper	sillimanite
barite	kaolin	silver
beryl	krennerite	sphalerite
biotite	kyanite	sphene
bismuth	lazulite	spodumene
calcite	magnetite	staurolite
cassiterite	mica	sulfur
cerussite	molybdenite	talc
chalcanthite	monazite	titanite
chalcedony	muscovite	topaz
chlorite	native copper	tourmaline
corundum	olivine	uraninite
covellite	opal	vermiculite
diaspore	orthoclase	xanthothane
dunite	plagioclase	zeolites
enargite	pyrite	zircon
epidote	pyrolusite	

Rocks are classified as igneous, sedimentary, or metamorphic—classifications that reflect the manner in which they were formed. An igneous rock is created from molten material, formed either deep within the crust of the earth (e.g., granite) or at or near the surface (e.g., lava or volcanic ash). Sedimentary rocks are formed by the transportation and deposition of sediments into water (e.g., sandstone and shale) or by the precipitation of minerals out of saturated water (e.g., limestone or gypsum). Metamorphic rocks are formed by forces of heat and/or pressure that change them chemically and structurally (e.g., schist or gneiss).

The igneous rocks of the Blue Ridge and Piedmont are found in two forms: (1) as intrusive granitic and gabbroic plutons, pegmatite, and diabase dikes and sills; and (2) as extrusive lavas and tuffs. The only sedimentary rocks found in the Piedmont area are fanglomerates, which are mixtures of sediments originally deposited in an alluvial fan and later consolidated into rock. These are found in the Crowburg Basin, a small Triassic rift basin at the Piedmont and Coastal Plain boundary in Chesterfield County. The vast majority of rocks in the Blue Ridge and Piedmont are metamorphic. They range from metamudstones, so lightly altered that trilobite fossils can be seen in them, to highly metamorphosed gneisses and amphibolites.

One key to deciphering the history of the Blue Ridge and the Piedmont is to find the degree of metamorphic change that the rocks in these regions have undergone. It is through both the microscopic and macroscopic study of the chemical and physical structures of the minerals and rocks that evidence may be gathered in order to determine what kind of environment caused these minerals and rocks to form. Because the atoms that make up the rock-forming minerals of the earth bond with each other according to very definite laws of chemistry and physics, their chemical bonding is predictable—based upon the original composition of the minerals and the amount of heat and pressure applied to them over time. Minerals do not occur randomly; rather, they arise as products of specific forces including chemical composition, pressure, and temperature. For example, some rocks in the Blue Ridge and Piedmont contain the mineral sillimanite, which attests to severe metamorphism. In contrast, others contain the mineral pyrite that is found in sulfide deposits that form in relatively low-temperature and low-pressure environments. The scale used to measure these mineral changes is called the metamorphic gradient.

An example of the metamorphic gradient is the effect of heat and pressure on shale. As medium heat and pressure are applied to the shale, its mineralogy alters. The shale becomes rich in muscovite mica, quartz, and plagioclase feldspar that forms a schist. With increasing heat and pressure, its mineral composition continues to change to include biotite, then garnet, staurolite, and kyanite. With the maximum heat and pressure, the original shale will become a coarsely foliated, sillimanite schist, then a gneiss. If the gneiss were heated beyond this point, it would melt and become a magma.

South Carolina mineral display (courtesy: Read Miner)

Some metamorphic rocks found in the Blue Ridge have been heated almost to the melting point due to continental collisions. Their minerals became so plastic under the extreme metamorphic conditions that they flowed. Evidence of this can be seen in many road cuts in the upstate. The gneiss of Whitewater Falls (just at the South Carolina/North Carolina border on S.C. Highway 130) is a good example. When metamorphic gneiss is injected with melted rock it is called a migmatite—another common rock of the Blue Ridge. Although the metamorphism that affected South Carolina's rocks took place millions of years ago, the rocks carry traces of the immense heat and pressure that formed them. Geologists can determine from the evidence presented by the rocks themselves which ones formed the core of the continental collision and which were located on the edges.

Geologists, much like detectives, study clues that allow them to investigate geological problems and offer hypotheses. Geological

Table 3.1 Common Rocks of the Blue Ridge and Piedmont

Igneous	Metamorphic
anorthosite	amphibolite
basalt	argillite
diabase	eclogite
diorite	gneiss
dunite	granulite
gabbro	hornfels
granite	marble
granodiorite	metabasalt
monzonite	metabreccia
norite	metaconglomerate
pegmatite	metadactite
peridotite	metagreywacke
syenite	metasandstone
tonalite	metasiltstone
	metavolcanics (tuffs, lapilli, rhyolite)
	migmatite
	mylonite
	schist (various types)
	serpentinite
	slate
	phyllite
	quartzite

structures found in the Blue Ridge and Piedmont—including small (almost microscopic) crenulations, huge synclines and anticlines, faults, plutons, thrust sheets, shear zones, mylonitic zones, domes, and windows—tell the story of the formation of the rocks that comprise them. Many of the clues needed are also found in the mineralogy and metamorphic grade of each rock, the relative positions of the rocks to each other, and the decay of the radioactive minerals found within them. This process of search and discovery is continuing today in South Carolina as geologists and geophysicists piece together the story of the southern Appalachians. Refer to the South Carolina Geologic Map in the color plate section for reference to rocks of the Blue Ridge and Piedmont.

THE BLUE RIDGE REGION

The oldest rocks in South Carolina are found at the base of the Blue Ridge Mountains. These are continental Precambrian Grenville basement rocks (see chapter 2) that have been highly metamorphosed to gneisses. They include the Toxaway and Wiley Gneisses and can be found in Oconee County off S.C. Highway 107 (see Hatcher, *Carolina Geological Society Field Trip Guidebook,* 1976). Other, younger gneisses, schists, amphibolites, metagreywackes, and pegmatites are also present. These rocks were previously the outer margin clastic sediments and volcanics that formed during the Precambrian rifting, creating the ancient Iapetus Ocean. These rocks were then metamorphosed to their present form due to continental collision. During the Taconic Orogeny, basalt dikes and flows were metamorphosed to amphibolite. The metamorphic grade decreases from the central Blue Ridge eastward toward South Carolina, implying that the core, or central portion, of the Blue Ridge is farther to the west in North Carolina. Although the Blue Ridge shows few effects of the later Acadian Orogeny, traces can be seen as crenulations in some rocks. The Blue Ridge also contains Mesozoic diabase dikes.

The Brevard Zone

The Brevard Zone is a one-third-to-two-mile-wide zone of highly broken, mylonitic rock that separates the Blue Ridge thrust sheet from the Inner Piedmont. It extends from Virginia to Alabama. In South Carolina the Brevard Zone runs along S.C. Highways 196, 107, and 130 and is seen in places as a valley trending northeast to southwest. Because the valley was created from fault movements, it contains rocks crushed at their contacts and weakened. This weakened condition makes it easier for water to break down the rocks further and thus accelerate their erosion. The rocks are highly deformed, mylonized schists and metasediments that, when picked up, can be broken and crushed by hand. The crushed rock found in the area testifies to the tremendous force and movement that created the fault. Also found in this fault zone are garnet schists and Jurassic diabase dikes.

The nature and origins of the Brevard Fault have been debated for many years. It does not represent the suture that was formed between North America and Africa during the

Mylonitic weathered schist in the Brevard Zone

Alleghenian Orogeny, as was thought for many years by some geologists. The fault instead appears to be the long-eroded root zone of a thrust sheet: the line along which great quantities of rock that formed the thrust sheet were shoved up toward the northwest and over other rock. In the process, this movement squeezed up some of the flat-lying sediments of the now-buried ancient continental shelf. Some of the exposed rock began as dolomitic limestone, but it was heated into marble as it was squeezed between the two highly metamorphic zones of the Blue Ridge and the Inner Piedmont. The presence of this altered carbonate in this particular environment indicates that it was forced up from far below the surface. It also further shows that the Blue Ridge and Piedmont are not native to the North American continent, but are allochthons (previously unattached fragments of land).

Streams flowing across the hard rock of the Blue Ridge and Inner Piedmont have produced many waterfalls. The highest series of falls in eastern North America is found at Whitewater Falls. There are, however, over fifty other waterfalls in the state, many of which are found in Oconee County. A full list of the state's waterfalls, which includes descriptions of each and their locations, is available from the South Carolina Department of Parks, Recreation, and Tourism (see the resources list).

Places to Visit in the Blue Ridge

Sumter National Forest—Andrew Pickens Ranger District, 112 Andrew Pickens Circle, Mountain Rest, South Carolina, 29664, 803-638-9568; in Oconee County—walking trails through forest with many rock exposures, especially along the Chatooga River and Whitewater Falls; includes Ellicott Park Wilderness.

Woodall Shoals on Chatooga River—U.S. Forest Service Road 757 off Oconee County Road 538 off U.S. Highway 76, in Oconee County—metamorphic rock sites and river scenery.

Oconee State Park—north of Walhalla on S.C. Highway 107—mountain and river scenery and many metamorphic and igneous rock sites.

Buzzard Roost—north of Walhalla in Oconee County off S.C. Highway 28—a 285-acre South Carolina Non-Game and Heritage Trust Preserve

property that contains a fine marble outcrop and harbors at least eight endangered species of plants.

Chatooga National Wild and Scenic River (in Sumter National Forest)—S.C. Highways 107 and 28; U.S. Highway 76—metamorphic rock
sites and lovely fifty-mile-long white water river scenery.

Brevard Fault Zone—S.C. Highways 196, 107, and 130—mylonite rock
can be seen in some parts of the fault zone.

Ellicott Rock Wilderness—contact USDA Forest Service, 112 Andrew
Pickens Circle, Mountain Rest, South Carolina 29664, 803-638-9568—sheer
rock faces and rare plant habitats.

Whitewater Falls—S.C. Highway 130 at North Carolina border, in
Oconee County—spectacular falls that begin in North Carolina and end in
South Carolina.

Yellow Branch Falls—S.C. Highway 107 north of Walhalla, in Oconee
County.

THE INNER PIEDMONT BELT

The Inner Piedmont belt is bounded on the northwest by the
Brevard Fault Zone, on the southeast by the Kings Mountain
belt, and along its southern extension by the Lowndesville Fault
Zone. Many geologists believe that this terrain is a continental
fragment pulled off the North American continent during Precambrian rifting, then reattached during the Taconic Orogeny.
Other geologists believe that the Inner Piedmont belt was not
formed in North America, but rather that it is an exotic terrain
(formed outside North America). The Inner Piedmont belt overlies Grenville basement rock, but none of this ancient North
American crust can be seen at the surface in South Carolina
because all traces of it were buried in the collision.

Formed from eroded thrust sheets, the Inner Piedmont belt is
composed of highly metamorphosed gneisses, schists, amphibolites, and some ultramafic bodies containing dunite and peridotite. The ultramafic rocks are thought to have originated as a
sea floor that had been scraped onto the continent during subduction and collision. The original material from which the
schists and gneisses were made—their protoliths—were both
volcanics and terrigenous sedimentary rocks. The Inner Piedmont

belt also contains Late Ordovician and Early Silurian granitic plutons that were emplaced at a depth of about 12.5 miles as the Inner Piedmont fragment collided with North America.

Because continental rocks "float" on top of denser rock, they maintain isostasy (or equilibrium). As mountain ranges erode, they rise like empty ships in the water and are further eroded. Because the uplift experienced by the Inner Piedmont belt is estimated to have been from 8 to 15 miles, granite that formed miles underground has been brought into view at the surface. Two of South Carolina's most visited exposed plutons are found in the Inner Piedmont belt: Caesar's Head (elevation 3,266 feet) and Table Rock (elevation 3,197 feet). Caesar's Head was formed about 409 million years ago, but it has taken many millions of years for it to rise toward the surface and weather into view. Both Table Rock and Caesar's Head stand out prominently as monadnocks because their granitic gneiss is more resistant to erosion than the schists, amphibolites, and metavolcanics of the surrounding countryside. Other interesting plutonic rocks of the Inner Piedmont include the Cherryville granite near Gaffney, the Cold Point pluton in Laurens County, and the Pacolet monzonite in Spartanburg County.

There are many large faults in the Inner Piedmont where interesting mineralization has occurred. The faults trend northeast to southwest and can be clearly located on topographic maps. One such fault is found along Gap Creek on U.S. Highway 25 north of Greenville. Evidence of both rock movement and the intrusion of zeolite minerals and epidote can be found along the road cut 2.5 miles south of the North Carolina border. The presence of zeolite minerals indicates a low-temperature, low-pressure environment rich in calcium, sodium, and potassium. These zeolite minerals formed from elements that were carried in 200–400 degrees Fahrenheit fluid and deposited in the cracks of the rock as the fluid cooled. They are found in small veins within the rock and include heulandite, chabazite, and stilbite. Also found along the cracks are quartz crystals, purple fluorite, prehnite, and the rare mineral, babingtonite. Drusy quartz and jasper can also be found in the Gap Creek area.

Caesar's Head,
Greenville County

Estatoee Waterfall (courtesy: South Carolina Department of Parks, Recreation, and Tourism, Tourism Division)

Caesar's Head, showing exfoliation of the granite

Fault near Gap Creek, U.S. Highway 25, Greenville
County (courtesy: Read Minor)

Inner Piedmont waterfalls in Pickens County include the Reedy Cove Falls and the Twin Falls, while those in Greenville County include Raven Cliff Falls and Rainbow Falls.

Places to Visit in the Inner Piedmont Belt

Eight state parks lie in the Inner Piedmont belt (see the South Carolina Department of Parks, Recreation, and Tourism *State Parks Guide*).

South Carolina Non-Game and Heritage Trust Preserves in the Inner Piedmont (write to the South Carolina Department of Natural Resources for a detailed description and map of each site—see the resources list):

Mountain Bridge Wilderness and Recreation Area—10,000-acre preserve in which lie Caesar's Head and Jones Gap State Parks, as well as several Heritage Trust Preserves.

Raven Cliff Falls—off S.C. Highway 11 north of Greenville near Caesar's Head State Park—scenic vistas of the Piedmont from a two-thousand-foot escarpment; part of the Mountain Bridge Wilderness and Recreation Area.

Ashmore Heritage Preserve—529 acres in Greenville County—contains an open seepage area on a steep granite exposure, with rare plants, and striking views of the Piedmont; part of Mountain Bridge Wilderness and Recreation Area.

Chandler Heritage Preserve—251 acres in Greenville County—contains a cataract bog that harbors rare plants; part of Mountain Bridge Wilderness and Recreation Area.

Watson Heritage Preserve—1,660 acres in Greenville County—a mountain bog containing rare plant species; part of Mountain Bridge Wilderness and Recreation Area.

Estatoee Creek Heritage Preserve—U.S. Highway 178 at Estatoee Creek, crossing north from S.C. Highway 11, 8 miles—steep mountain gorge featuring an old-growth hemlock forest and several rare plants.

Glassy Mountain Heritage Preserve—65 acres in Pickens County, off S.C. Highway 183 east of Pickens—granite monadnock that contains rare plant communities and hiking trails; a clear view of the Piedmont and Blue Ridge Mountains at the top.

Furman University—U.S. Highway 25 in Greenville—rock and mineral display.

Clemson University—rock, mineral and fossil display.

Cherokee Foothills Scenic Highway—S.C. Highway 11—road to Caesar's Head, Table Rock, Pinnacle Mountain, and many other mountain sites; many waterfalls visible.

Sassafras Mountain—from Pickens, north on U.S. Highway 178, then Pickens County Road 199—highest mountain in South Carolina.

Paris Mountain Monadnock and Nature Center—off S.C. Highway 253 north of Greenville in Greenville County.

Roper Mountain Science Center—off U.S. Interstate 385 in Greenville, 501 Roper Mountain Road—fossils and rocks; educational programs.

Gap Creek Fault—U.S. Highway 25 north of Greenville near South Carolina/North Carolina state line—zeolites and fluorite in rock faces along fault zone.

THE KINGS MOUNTAIN BELT

The Kings Mountain belt is a highly mineralized suture zone (or boundary) between the Inner Piedmont continental fragment and the Charlotte belt and the Carolina slate belt. Many of the rocks of the Kings Mountain belt are similar to those found in the Charlotte belt and the Carolina slate belt. Geologists believe that although the three belts compose the same terrain, their geological differences are caused by the degree of metamorphism that each area underwent as it was welded onto North America. The Kings Mountain Shear Zone is 340 miles long and runs from Alabama to North Carolina. Unlike the high-grade metamorphic rocks of the Inner Piedmont (schists and gneisses), the rocks of the King's Mountain Shear Zone include lower-grade metavolcanics and metasediments. Because this area underwent a lower-grade metamorphism, its minerals are more stable at the surface and take longer to chemically weather. As a result, the topography of this narrow belt is high. Rocks found here include phyllites, schists, quartzite, metaconglomerates, and marble; minerals include aquamarine, lithium, and gold.

In South Carolina along the north/northwest boundary of the Kings Mountain Shear Zone, where the force of impact was most intense during collision, the rocks are highly deformed. Some oceanic rock caught up in the collision melted to become peridotite, and metamorphosed into amphibolite and serpentinite.

At this boundary, mylonitic rocks (contact rocks crushed by faulting) are also found. The metamorphism of the collision turned sandstone into quartzite and limestone into marble, and created granitic intrusives that today are seen as resistant monadnocks, ridges, and knobs that dot the landscape. For a detailed description of the geology of this area, see: *Geological Investigations of the Kings Mountain Belt and Adjacent Areas in the Carolinas* published by the Carolina Geological Society.

Places to Visit in the Kings Mountain Belt

Limestone College Marble Quarry—in Gaffney, U.S. Highway 29 in Cherokee County.

Kings Mountain State Park—S.C. Highway 216 off U.S. Interstate 85 north of Gaffney—many rock outcrops.

Smyrna—many old mines and diggings north and south of the town in York County—seek permission before entering private property.

Soapstone Ridge—near Pacolet Mills—a metamorphic ridge composed of varying amounts of talc, mica, chlorite, and other minerals; amethyst also found in the area.

THE CHARLOTTE BELT

The Charlotte belt is the most highly intruded of the three belts making up the Carolina terrain. Its plutonics are deeper and older than those found in either of the other two belts. Many geologists believe that the Charlotte belt represents the central core of the island arc that collided with North America, and that it is the volcanic source of the tuffs (now metamorphosed) that are found over the entire area in the Kings Mountain belt, Charlotte belt, and Carolina slate belt. The Charlotte belt represents the axis (or heart) of a large syncline (or synform) with the Kings Mountain belt to the northwest and the Carolina slate belt to the south and southeast, representing its limbs.

The presence of serpentinite in South Carolina once convinced some geologists that the Charlotte belt was a mélange. A mélange is a heterogenous medley of rock materials created as ocean bottom is scraped off against the edge of a continent as the ocean bottom and continent move toward each other. An example of this activity is found along the northern coast of California where the North American plate overrode the Pacific plate during the Middle Jurassic to Early Cretaceous Periods. Evidence has shown, however, that the Charlotte belt is actually the core of an island arc. Additional evidence of an island arc is provided by chemical analyses of the rocks and minerals found in the area. The rocks and minerals of this belt are very similar in composition to those of a more recent island arc, the Lesser Antilles, which has been built from a subduction zone now active in the eastern Caribbean.

Granitic and gabbroic plutons, ranging in age from 285 to 735 million years old, are found in the Charlotte belt. This belt contains such mafic rocks as metagabbro, amphibolite, greenstone-metabasalt, and the ultramafic rock serpentinite. The Charlotte belt contains rocks of high metamorphic grade and complexity. Good rock collecting sites include Newberry County—the Newberry granite: a quartz monzonite (adamellite) with biotite gneiss inclusions; Laurens County—a metagranite/gneiss complex near the town of Gray Court; Laurens–Spartanburg County line—porphyritic granite with large intrusions of biotite schist and gneiss; Spartanburg and Union Counties—vermiculite-bearing schists between the towns of Woodruff and Cross Keys; and Union County—the one hundred-square-mile Bald Rock granite porphyry, the largest postmetamorphic batholith in the state. A detailed field trip guide to the rocks of the Charlotte belt and the Carolina slate belt is *Granitic Plutons of the Central and Eastern Piedmont of South Carolina* by Wagener and Howell.

Places to Visit in the Charlotte Belt

Sumter National Forest—in Union and Chester Counties.
Rose Hill Plantation State Park—south of the town of Union, views of the Tyger River on Sardis Road, off U.S. Highway 176.

Bald Rock batholith—Union and Cherokee Counties—100 square miles of granite and quartz monzonite that lies north of the town of Union; outcrops are visible.

Rock Hill—in Martin Marietta Quarry, off U.S. Interstate 77—gabbro pluton underlies the area; outcrops.

Catawba Nuclear Station—near Fort Mill on U.S. Interstate 77—Energy Quest public display.

Gold Hill Mines—off U.S. Interstate 77 at Fort Mill, Carowinds Exit onto U.S. Highway 21 (business)—gemstones and gold panning for a fee.

The Museum—U.S. Highway 25 in Greenwood—rocks and minerals.

Museum of York County—Rock Hill, 4621 Mount Gallant Road off S.C. Highway 274—rocks, minerals, and fossils.

THE CAROLINA SLATE BELT

The Carolina slate belt, which extends 600 miles from Georgia to Virginia, is composed predominantly of rocks that reflect slight- to medium-grade metamorphism. The andesitic volcanic ash that blew out of volcanoes in the Charlotte belt 450 million years ago can still be clearly seen in the argillites and schists of this area. Good examples of argillites are found along S.C. Highway 207, between Pageland and Chesterfield. Extensive examples of schists can be seen in the Haile Gold Mine tailings along U.S. Highway 601, just north of Kershaw. These metatuffs are banded with iron, and the sediments are only mildly altered to form a sericite schist (dated at 465 million years). The "slates" of the belt are more accurately designated as argillites or phyllites: rocks that were formed from assorted parent sediments, including volcanic ash and offshore muds. Argillites are clay-rich rocks, metamorphosed to a very low level. Phyllites are intermediate metamorphic rocks, between slate and mica schist. Good examples can be seen along S.C. Highway 34, east of Ridgeway.

In the southern portion of the Carolina slate belt, just northwest of Columbia, sediments can be found that are not volcanic, but are lightly metamorphosed mudstones and sandstones from the trailing edge of the island arc. Middle Cambrian-age trilobites were first discovered in 1982 in these mudstones (the

Asbill Pond Formation). Their discovery helped to confirm the nature and origin of the Carolina slate belt, for these trilobites included nine genera that were native to what is today Bohemia, in southern Europe.

Slate belt deposits of metavolcanics near Ridgeway Gold Mine in Fairfield County

The Carolina slate belt contains many shallow plutons that were formed during the Alleghenian Orogeny. They range in thickness from ½ to 9 miles. The force of the African–North American collision in the Permian Period was so great that rocks at depth melted to form relatively shallow granitic plutons. It has been estimated that the depth of greatest metamorphism in the Blue Ridge was 13.5 miles. In comparison, many of the Piedmont plutons were formed at only about 5 to 7.5 miles below the surface. A partial list of Alleghenian age granitic plutons found in the Carolina slate belt follows.

Table 3.2   Alleghenian-Age Granitic Plutons Found in the Carolina Slate Belt

Pluton	Age (in years)
Columbia granite	286 million
Lake Murray granite	311 million
Lexington granite	293 million
Winnsboro granite	295 million
Pageland granite	297 million
Liberty Hill granite	295 million

These shallow plutons lie in a diagonal line across the Piedmont from Pageland southwest through Liberty Hill and Winnsboro, to the Saluda County line. They include the largest flat rock granite exposure in South Carolina: Forty Acre Rock in Lancaster County. This outcrop (14 acres) is composed of the coarse-grained Pageland granite that intruded into the metasediments and metavolcanics of the area, creating aureoles of hornfels around the pluton, visible at road cuts (see Wagener and Howell). Fractures, fissures, and small depressions

Forty Acre Rock in Lancaster County. Shows a solution vernal pool in which plants and animals live through part of the year (courtesy: South Carolina Department of Natural Resources, Non-Game and Heritage Trust)

then formed in the rock as the granite of Forty Acre Rock weathered. These depressions, called solution vernal pools, catch rainwater and form habitats for flat rock plants (including the rare pool sprite). Various sedum, lichens, and mosses have adapted to withstand the heat and dryness of the rock face for most of the year. This unusual ecosystem is now protected as a Heritage Trust Preserve.

The Liberty Hill granite pluton lies southwest of Forty Acre Rock. It is largely grey and finer grained than the Pageland granite, which it intrudes. It is visible in many road cuts between Heath Springs and Liberty Hill, and it forms an interesting boulder-strewn landscape, including Hanging Rock south of Heath Springs.

Hanging Rock, Lancaster County

The Winnsboro area contains a blue granite that is a particularly fine-grained quartz monzonite chosen as the South Carolina state stone—the Rion adamellite. Because the stone has very few flow features that would mar its texture, it has been in high demand for monument stone throughout the region. It has been used for decades as the building stone for many of the houses and other buildings in the Winnsboro area. Exposed in many sites, Rion adamellite can be found in the inactive Rion Quarry off S.C. Highway 269 south of Winnsboro.

Little Mountain in Newberry County, the highest point of land between Charleston and Greenville, is an example of a Carolina slate belt monadnock. The monadnock rises 275 feet above

Winnsboro blue granite used in buildings and fences such as the Greenbriar School, near Rion in Fairfield County

the Piedmont and 825 feet above sea level. It is made up of schists and resistant kyanite quartzite that are bounded on the north by granites and gneisses. Little Mountain has a complex mineralogy that includes kyanite, pyrophyllite, pyrite, rutile, ilmenite, hematite, lazulite, corundum, cassiterite, and gold.

There is a series of Triassic red bed rift basins that run from South Carolina north to New Jersey. These rift basins were formed as Africa and North America began to separate during

Crowburg Basin, .9 mile east of Crowburg on S.C. Highway 207. Road cut contains red beds of fanglomerates and sedimentary deposits at edge of the Triassic basin

the mid-Triassic, producing a series of large grabens as the extending crust thinned and weakened. South Carolina contains the southernmost exposure of a Triassic red bed rift basin at the surface: the Crowburg Basin in Chesterfield County. This basin, related to the larger Wadesboro Basin in North Carolina, is a wedge-shaped depression approximately 6 miles long and almost 2 miles wide. The escarpments that rim the Crowburg Basin rise from 50 to 100 feet making them visible to the passerby. This basin is located northwest of Pageland along S.C. Highway 207 and running through the village crossroads of Crowburg. South Carolina contains other rift basins, but these basins lie buried under the sediments of the Coastal Plain. The

Crowburg Basin contains red fanglomerates—sediments that are composed of heterogeneous materials originally deposited as an alluvial fan, but later cemented into solid rock. These sediments are highly weathered, but the broken pieces of argillites (clay-rich shales) can be easily identified. These argillites, derived from volcanic ash, are interbedded with maroon to red shale and siltstone with some vein quartz.

Mesozoic diabase dikes, which were formed about 80 to 100 million years after the Alleghenian Orogeny, are prevalent in the Carolina slate belt. These dikes were emplaced as Africa and North America separated during the Mesozoic. The dikes trend largely north and northeast and run from offshore, through South Carolina to North Carolina. The Martin Marietta Granite Quarry in Cayce contains a very visible diabase dike, which trends to the northeast and cuts through the older Pennsylvanian granite that underlies the Columbia area. The largest diabase dike in the eastern United States is found in South Carolina: the Flat Creek Dike (or "Great Dike"), north of Kershaw in Lancaster County on U.S. Highway 601. This dike, over 1,000 feet across in places, continues into North Carolina.

Places to Visit in the Carolina Slate Belt

Flat Creek Heritage Preserve—Forty Acre Rock, off U.S. Highway 601 on Lancaster County Road 27 near Taxahaw—largest flat rock granite in the state.

Anderson Granite Quarry—S.C. Highway 269 near Rion in Fairfield County—pink pegmatite dikes running through the blue granite.

Martin Marietta Granite Quarry—in Cayce in Lexington County—contains diabase dike.

Brewer Gold Mine Area—near Jefferson on S.C. Highway 265 in Chesterfield County.

Barite Hill Gold Mine Area—7 miles south of McCormick on S.C. Highway 28 in McCormick County.

Haile Gold Mine Area—on U.S. Highway 601 north of Kershaw in Lancaster County—large tailing piles.

Tarmac Granite Quarry—Olympia Mill district in Columbia in Richland County—field trips for school groups. Call for appointment: 803-771-0090.

Crowburg Basin—S.C. Highway 207 northwest of Pageland in Chesterfield County—Triassic basin.

Fall Zone on the Saluda and Broad Rivers—in Columbia, view near the Riverbanks Zoo on U.S. Interstate 126 or from the Broad River Road bridge just west of the Greystone Boulevard/Broad River Road intersection.

Hanging Rock—take S.C. Highway 15 to Hanging Rock Road, 2.5 miles south of Heath Springs in Lancaster County—Lancaster County granite boulders.

Molly's Rock Recreational Area—Santee National Forest on U.S. Highway 176 in Newberry County.

Lake Murray Spillway Area—S.C. Highway 6 in Lexington County—gneiss, garnets.

Little Mountain—U.S. Highway 76 in Newberry County—kyanite in quartzite; monadnock.

Ninety-Six Southern Brick Company—P.O. Box 208, Ninety-Six, South Carolina 29666, 803-543-3211, in Richland County—field trips for school groups; call for appointment.

South Carolina State Museum—Gervais Street in Columbia in Richland County—rocks, minerals, fossils, resources, geological history of South Carolina display.

McKissick Museum—University of South Carolina in Columbia in Richland County—large collection of minerals.

THE KIOKEE AND BELAIR BELTS

Two small belts lie at the southwest boundary of the Carolina slate belt—the Belair and the Kiokee. They both continue into Georgia and can be seen in the Augusta area along the Savannah River. The rocks of the Belair and Kiokee belts have great similarities to those of the Carolina slate belt, and are considered by many geologists to be part of that belt. Both Kiokee and Belair belt rocks include low-grade metavolcanics and metasediments, but the rocks of the Kiokee belt are further metamorphosed and deformed. They include granitic intrusions, diabase dikes, quartzites, quartz-mica schists, amphibolites, serpentinite, and hornblende gneiss. The Kiokee belt is thought to be a subduction-related complex of rocks that was accreted to the continent during the Taconic Orogeny and was later highly faulted during the Alleghenian Orogeny. On the north, the

Kiokee belt is bounded by the inactive Modoc Fault Zone that runs west from Lake Murray, near Irmo, to Warrenton, Georgia.

PIEDMONT SOILS

The Piedmont soils of South Carolina, produced by the wearing down of Blue Ridge and Piedmont rocks, are of both geological and historical interest. Geologically, these soils illustrate classic weathering and erosional processes that act to break down mountains. Historically, they played a key role in the establishment of America's soil conservation program.

Saprolite soil near Peak, Newberry County

The thin mountain soils of the Blue Ridge quickly give way to the deep-to-moderately deep and well-drained soils of the Piedmont. The surface layers contain sandy loam, while the subsoils range from sandy clay loam to clay. Saprolite soils—earthy, clay-rich soils with thoroughly decomposed rock formed in place by chemical weathering of igneous or metamorphic rocks—are common in the Piedmont. Saprolite soils predominate in the interstream areas and are notably red, colored by the iron-rich minerals weathered from the original rocks. The weathering that forms these soils takes place as ground water moves within the rock, removing as much as 60 percent of its mass while maintaining its volume. The rock literally rots in place. As a result, most saprolites contain from 10 to 25 percent clay, with some containing as little as 3 percent and others as much as 70 percent. One meter of saprolite forms in about 250,000 years, which is a relatively rapid rate of weathering. Exposures of saprolite in South Carolina range in age from Miocene to Pleistocene. The saprolites erode easily and fall apart readily when disturbed, often exposing quartz veins. One such site is near the town of Peak, in Newberry County.

During the early twentieth century, the soils of the Piedmont contributed to a greater understanding by farmers and the public of the fragile nature of topsoil. This knowledge was gained, however, at a tremendous price. As cotton boomed during the nineteenth and early twentieth centuries, these upland soils were farmed extensively—little thought was given to soil protection. Because agriculture was unrestricted and many farmers did not understand how best to guard their soils against erosion, a great deal of damage was done to the land. The clear-cutting of trees, the overproduction of cotton, and the failure of farmers to use crop rotation, fallow field practices, and contour plowing rapidly depleted the soil. Deep gully erosion occurred as the steep slopes of the Piedmont continued to be ignored by the farmers. Today, the South Carolina Piedmont region is considered one of the most severely eroded agricultural areas in America, with an average erosional depth of 9.5 inches. Scientists have estimated that up to 40 percent of South Carolina's lands were ruined through these early agriculture practices. The tremendously high rate of erosion in the area also adversely affected the rivers in the upstate, for they became inundated with sediments that created swampy bottomlands. Streams and rivers flowed with red mud harming the wildlife, including fish such as trout.

The United States Soil Conservation Service was established in 1933 to help repair and slow the effects of erosion nationwide. The Service helped plant trees on damaged lands, and taught farmers to use wiser farming practices in order to regenerate and preserve the irreplaceable topsoils. The huge gullies that formed in the Piedmont are now largely reforested with longleaf pine trees. The site of the first erosion control project in the southeast was Berry's Gully in the Poplar Springs community of Spartanburg County, between Woodruff and Spartanburg.

J. L. Berry's Farm, with a gully, showing severe erosion, November 1936, in Poplar Springs community between Spartanburg and Woodruff (courtesy: U.S. Soil Conservation Service)

Berry's Gully after erosion was brought under control, 1930s; site of first erosion control work in the southeast (courtesy: U.S. Soil Conservation Service)

Extensive erosion near Inman, Spartanburg County,
in the 1930s (courtesy: U.S. Soil Conservation Service)

MINERAL RESOURCES OF THE BLUE
RIDGE AND PIEDMONT

The Blue Ridge and the various belts of the Piedmont contribute distinctly to South Carolina's economy. Because of their
differing geology, they provide some of the most valuable and
rare minerals found in the state (including gold and silver) and
produce some of the most mundane resources as well (such as
crushed granite and sand). See the South Carolina Mining Map
in chapter 5 for more information on minerals found in the state.

Although the Blue Ridge is a source of mineral wealth in Georgia, North Carolina, and Virginia, in South Carolina it produces mainly crushed stone. Gold was found in the Blue Ridge region of South Carolina in the nineteenth century, but there are no active mines there today. Rather than producing great mineral wealth, the natural beauty of this area has made it most valuable as a tourist site. Almost all of the Blue Ridge Mountains situated in South Carolina lie in the Sumter National Forest. It is an ideal place for camping, exploring, and rockhounding, and offers a landscape that includes fast-flowing rivers and waterfalls, quiet lakes, and spectacular mountain views.

The Inner Piedmont belt produces granite as crushed stone, which is quarried in Pickens, Greenville, and Spartanburg Counties. Sand is also mined from river deposits. The economically important mineral, vermiculite, is surface-mined in several localities in the state—mostly in Laurens and Spartanburg Counties. Vermiculite is an altered mica found in gneisses and schists that can expand twenty-six times its volume when heated. It is used extensively as insulation, as a filler for both plaster and fire-retardant wallboard, and as an indispensable ingredient in water-retaining potting soil. South Carolina ranks first in the nation in the production of this important and versatile mineral. Granites and gneisses found in Greenville and Spartanburg Counties often include the heavy mineral monazite (which contains the rare elements thorium, cerium, and lanthanum). Historically, monazite was an important mineral in South Carolina—identified in the state in 1880 and mined until 1918 as a source of thorium for use in making mantles for gas lamps. Other heavy minerals such as ilmenite and rutile were also once mined in alluvial deposits, but their exploitation is no longer economical. The mineral lithium is found within spodumene in pegmatites along the Inner Piedmont/Kings Mountain belt boundary in Cherokee and York Counties, and was once mined there. Today, it is economically mined just across the state line in Kings Mountain, North Carolina.

The Kings Mountain belt has produced a very valuable assortment of minerals and rocks—including gold, barite, kyanite, mica, pyrophyllite, pyrite, iron, manganese, lead, silver, marble, and crushed stone. This region is an interesting area to explore for minerals, because the intrusive and metamorphic history of the Kings Mountain belt has produced a large variety of minerals important to the collector. The state gem, amethyst, is found near Pacolet Mills. Although some gem quality amethyst has been found in South Carolina, its occurrence is rare. The Smyrna area in York County was one of the sites of the nineteenth-century Carolina gold rush. Many old diggings are found throughout the quartz diorite pluton northeast and southwest of the town. Barite, iron, manganese, graphite, and marble are found northwest of Smyrna, and kyanite is found to the north. A substantial barite mining industry existed near Kings Creek from 1885 until 1966. Barite was found in schists and quartzites, often with tourmaline, chlorite, galena, sphalerite, chalcopyrite, pyrite, and calcite. In the past small amounts of tin have been mined near Gaffney. The Kings Mountain belt was the site of South Carolina's Old Iron District, which produced a great deal of iron both as an export to nearby states and as a major source of domestic iron (from colonial times after 1755 to the end of the nineteenth century). An important iron mine of the area was the Cameron Mine on Limestone Creek, near Gaffney. The iron found in the state is in the form of hematite and limonite. The minerals galena, chalcopyrite, and pyrite are often found with iron in this area.

During the 1800s and early 1900s, the Charlotte belt produced some gold, along with iron, silver, and copper in the mines of York County. Although gold exploration continues today, the belt now produces mainly sand, clay, granite, and gravel. Cherokee County produces manganese schist, which is used in brick production. Corundum was mined briefly in York County during World War II when—because of its great hardness—it was used to grind plates for radar and radio sets.

The Carolina slate belt produces the greatest quantities of two of the highest-valued rock and mineral commodities in the state: crushed stone and gold. Together these two products to-

Great Falls metagranite intruded by mafic dike, both weathered severely, S.C. Highway 97 north of Liberty Hill

taled almost one-half of South Carolina's mineral income in 1993. South Carolina's two active gold mines and three of its largest granite quarries are all located in the Carolina slate belt. Granite has been quarried in the state since 1786, and today South Carolina has thirty-four active granite quarries.

Other mineral resources in the Carolina slate belt include sand, kaolin, sericite, and argillite. Sericite, which is a white, fine-grained mica formed in metavolcanics, is mined in six sites in this belt (mainly in Kershaw and Lancaster Counties). It is used in the manufacture of electrical equipment, paints, cosmetics, and grouting. Argillite, one of the chief ingredients in the manufacture of brick, is mined in several active quarries in this belt.

Tarmac Granite Quarry in Columbia

Harpers New Monthly Magazine, August 1857: "Finding Gold"

CHAPTER 4

CAROLINA GOLD

California is usually thought of as the site of America's first gold rush; however, the Carolina gold rush began long before the 1848 historic discovery of placer gold deposits at Sutter's Mill in California. America's first rush for gold began in 1799 with the discovery of a seventeen-pound gold nugget in Cabbarus County, North Carolina. Quite by accident, Conrad Reed, the young son of a German immigrant farmer named John Reed, found the gold while he played in a streambed on his father's property. For 3 years, the true nature of the nugget remained hidden and served the family only as a doorstop! A Fayetteville jeweler later gave the geologically naive Reed $3.50 for the nugget and Reed considered himself lucky for the trade. He later discovered, however, that his doorstop had been sold by the jeweler for thousands of dollars. Reed quickly returned to Fayetteville and demanded that the jeweler pay him an honest price for the gold. The account of the famous nugget spread far and wide, and became the starting gun for America's earliest gold rush.

The first flush of gold exploration sent farmers into the rivers and hills in North Carolina to search out gold in streambeds and surface quartz veins. Before long, the excitement spread to South Carolina as well, where in 1802 gold was first discovered in Greenville County. Then, in 1827, a plantation owner named Benjamin Haile found a large nugget glittering in a stream bottom on his land north of Kershaw, in Lancaster

County. Unlike Reed, it did not take Haile 3 years to realize that he was holding gold. He began to work his placer deposits almost immediately, and rented out much of his 2,000 acres, in fifty-square-foot plots, to miners who became gold leasers.

South Carolina gold production began officially with Haile's first gold shipment to the Philadelphia mint in 1829. The Haile Mine became the richest single mine in the entire eastern United States—producing over $6 million in gold. The success of Haile's mine inspired the nineteenth-century search for gold throughout the South (including North Carolina and Georgia) and later in the West, and helped build fortunes for many. The majority of prospectors, however, left the gold fields with little to show for their work—for veins were quickly played out, the digging was hard, and the work ultimately proved too unrewarding and discouraging for most.

South Carolina experienced three significant periods of gold production with each interrupted by a major war. The earliest

Harpers New Monthly Magazine, August 1857: "Rocking Cradles"

period began with the startup of the Haile Mine in 1829, and lasted until the approach of the Civil War in 1861. Although records were not kept during the frantic days of early search

Hanging wall of the Haile Pit, Haile Mine, Lancaster County, 1906
(credit: L. C. Graton, no. 21, U.S. Geological Survey)

and discovery, official production figures from 1829 to 1858
state that 67,969 troy ounces of gold were produced in South
Carolina (at the time worth $1,404,960). It is estimated, how-
ever, that only about one-half of the gold production was ever

reported. Mints were established in 1838 in both Charlotte, North Carolina, and Dahlonega, Georgia—the two states where gold production was the highest. These two new mints facilitated delivery of the gold to the government because the gold could be collected and minted locally, avoiding the risk of a long and dangerous journey to Philadelphia.

Before the Civil War, some of the Carolina gold was coined by local craftspeople, including Christopher Bechtler, in North Carolina. From Rutherford County, Bechtler privately produced gold coins from 1830 until 1852, beginning with $2.50 and $5 gold pieces. Then, in 1832, he began to coin America's first gold dollars. From January 1831 until February 1840, Bechtler produced $2,241,850.50 worth of American gold coins. These coins provided a handy medium of exchange in the mining fields. Because the quality of Bechtler's work was so fine, and because currency was so needed in the United States, the federal government did not interfere. The government began to produce its own series of gold coins in 1849. Not surprisingly, most of the Bechtler coins disappeared; as the price of gold fluctuated, the coins were melted down because raw gold brought a higher price. The few Bechtler coins that survived are today worth a great deal to coin collectors.

The second profitable gold mine to develop in the state was the Brewer Mine near the town of Jefferson, in Chesterfield County. In addition to gold, the mine produced the copper minerals enargite and covellite and had served as a copper mine before the Revolutionary War. If gold had been discovered as early as the eighteenth century, however, no one spoke of it— gold production did not begin until 1828. At its peak, the Brewer Gold Mine employed from one hundred to two hundred workers. Like Benjamin Haile, the mine's owners leased out much of their property to miners, but in tiny twelve-square-foot plots. By 1838, production at both the Brewer and the Haile Mines began to sharply decline. Each mine struggled to retrieve the deeper gold, but higher production costs and the discovery of gold in California in 1848 ultimately eclipsed both of them.

The last great mine developed during the nineteenth century was the Dorn Mine in McCormick County, which began operation in 1852. This mine produced as much as $9,000 a day for its flamboyant owner, William Dorn. Dorn had searched for 15 years for gold in the McCormick area, striking it rich only 3 days before his property was to be sold to pay for his mounting debt. The mine made him a millionaire, and he lavished his wealth on his family and friends. During the Civil War, Dorn bought supplies for the Confederate Army. Ironically, after Dorn sold his property to Cyrus McCormick in 1874, the mine never again produced the quantities of gold it had during its glory days. Local legend has it that Dorn buried a fortune in gold somewhere near his home before he died, but the gold has never been found. Today the silent tunnels and digs of "Billy" Dorn's mine still run under and through the town of McCormick.

From 1820 to 1850, there were three hundred active gold mines in South Carolina. By 1859, only fifty-eight remained: twenty-one in Chesterfield and Lancaster Counties; nineteen in Spartanburg, Union, and York Counties; ten in Abbeville and Edgefield Counties; and eight in Greenville and Pickens Counties. Most of the mines worked underground lodes because the surface veins had been exhausted during the early heyday of the local gold rushes. When the technology necessary to retrieve the deep gold deposits became more complex, many of the mines did not last long.

At the beginning of the Carolina gold rush, the surface and near-surface gold was picked up, dug up, and panned relatively easily. A technique called mercury amalgamation was the most common process used by the miners to extract the gold at that time. Stream sediments, or clays containing gold, were panned or run through a sluice—a trough with mercury (called quicksilver) in grooves at the bottom. The gold bonded with the mercury that was then heated and burned off as a dangerous gas, leaving just the gold. After much of the surface gold had been mined, more complicated and expensive chemical processes were employed—primarily the cyanide-leaching process, which could refine even the very low-grade ores.

Carolina gold production was abruptly halted in 1861 by the Civil War. The early prospectors had been primarily farmers and slaves. The use of slave labor in the gold mines had allowed landowners, like Dorn, to make tremendous profits. For example, during the first sixteen months of its operation, the Dorn Mine extracted over $300,000 in gold at a cost of only $1,200—a profitable mine, indeed! Like the rice and cotton plantations, the gold mines relied on the doomed slavery-based economic system that, by the middle of the nineteenth century, was nearing its end.

During the Civil War most mines were deserted or shut down, but the Haile Mine was successfully converted to copper production and served as an important resource for the Confederacy. In 1865, however, after his memorable march through Columbia, General Sherman sent a small contingent of troops to Kershaw to burn and destroy the Haile Mine. Recovery from the mass destruction of the war was made very difficult by the severe labor shortage, and capital investment in mining was hard to find during the next two decades. Gold production in South Carolina fell to $10,000 per year.

After 1880, because of economic recovery and technological improvements, the lower-grade ores began to be processed profitably. In 1888, the Thies chlorination process was developed at the Haile Mine, which allowed gold to be more easily refined. The mine again became the South's leading gold producer. A boiler explosion in 1908, however, wrecked the mine, killed the manager, and stopped production until 1934.

In the late nineteenth century, the area surrounding the towns of Smyrna and Hickory Grove became South Carolina's second most productive gold mining area. Scores of mines and diggings were operated there, many of which produced gold, iron, copper, and silver. One of the most profitable mines was the Bar-Kat Mine (located 3 miles west of Smyrna on a small stream called Beech Branch). The arrival of World War I, however, stopped most of the mining, as people were called away to service and capital was diverted to wartime needs.

Harpers New Monthly Magazine, August 1857: "View of the Gold Hill Works"

Harpers New Monthly Magazine, August 1857: slaves in a mine boring into the rock in search of gold.

The third period of South Carolina gold production lasted from 1931 until 1943—driven by the 40 percent increase in the price of gold (which rose from $25 to $35 per ounce during the Depression). Many people who were out of work and financially desperate tried their luck in the Carolina gold fields, but most of them did not stay long. Many miners were attracted instead to the larger mines in the Dakotas, Colorado, and Alaska. Then, as World War II began, gold mining was forced to stop once again. In 1942, President Roosevelt ordered the nation to mine only those metals critical to the war effort, such as iron and tin. This was a death blow to most gold mining in South Carolina until the 1980s when renewed efforts to tap the gold deposits surfaced.

Shaft of the Ross Mine, Gaffney, Cherokee County, 1934 (credit: L. C. Graton, no. 22, U.S. Geological Survey)

CURRENT SOUTH CAROLINA GOLD PRODUCTION

The gold deposits of South Carolina are part of the Appalachian gold belt, which contains 1,400 gold-rich areas extending from central Alabama north to the District of Columbia. South Carolina's 130 gold mining areas are found primarily within the Carolina slate and Kings Mountain belts; however, there are deposits of gold in the Blue Ridge region and in the Charlotte belt as well. Companies from Utah, Texas, and Australia have continued their search for new gold within the Carolina slate belt along a strip that cuts diagonally from North Carolina to Georgia through parts of Chesterfield, Lancaster, Fairfield, Richland, Newberry, Saluda, and McCormick Counties. The state's active mines continue to be reevaluated for additional deposits as well.

Until recently, there were four active gold mines in South Carolina. Today, there are only two—the Ridgeway and Haile Mines. These mines have found considerable amounts of gold, making South Carolina the only gold-producing state east of the Mississippi River. As of 1992, the state ranked seventh in national gold production.

South Carolina's Active Mines

Kennecott-Ridgeway Mining Corporation: The Ridgeway Mine—located in Fairfield County—first gold poured: December 1988; production 1988–1994: 850,000 troy ounces.

Piedmont Mining and AMAX: The Haile Mine—located in Lancaster County—presently in a hiatus for exploration and land reclamation; first gold poured: 1985; production 1985–1994: 85,000 troy ounces.

Recently Closed Mines

Brewer Gold Company: The Brewer Mine—located in Chesterfield County near Jefferson—first gold poured: July 1987; production 1989–1992: 170,000 troy ounces; mining operation closed 1992; land reclamation efforts in progress.

The Nevada Gold Fields Company: The Barite Hill Mine—located in McCormick County—first gold poured: 1991; production 1991–1994: 60,000 troy ounces; mining operation closed 1994; land reclamation efforts in progress.

In 1995, the price of gold was about $380 per troy ounce, so the incentive to explore and the ability to efficiently work the low-grade gold deposits are keener than in the past. The costs of production are high, however, ranging from about $223 to $290 per troy ounce, so there is not a large margin for error in cost projection if a profit is to be made.

All of the gold mines in the state work low-grade ores where tremendous amounts of rock are needed to produce one ounce of gold. The Ridgeway Mine, for example, processes 31 tons of ore to retrieve a single troy ounce of gold. The mine uses the cyanide-leaching process to chemically extract the metal. Cyanide is one of the few substances capable of leaching gold. In this process, the gold ore is first crushed to a fine powder, then mixed with sodium cyanide, water, and quick lime to form a souplike slurry. Carbon, which is naturally attracted to gold, is later added to the slurry and atoms of microscopic gold then bond to the porous carbon. The gold is stripped from the carbon and further separated with a hydrogen chloride (HCl) solution and a concentrated 2 percent cyanide solution. Then carbon is again added and adheres to the gold. Finally, the gold is removed through electroplating. After being melted in a furnace,

The Kennecott-Ridgeway Mine, Fairfield County, looking south over the north and south pits on the 2,000-acre site; note pond in upper right (credit: Kennecott-Ridgeway Mining Corporation)

the gold is poured into 800-ounce bars, called dorés, which are then sent out of state for final refinement. Silver is present with the gold and is sold as a by-product.

Chemical by-products from gold ore processing, if not managed properly, can harm surface and underground water sources as well as wildlife. The Ridgeway Mine follows the most stringent guidelines ever imposed by a state or accepted by a mining company in the United States. In order to protect the environment, the Kennecott-Ridgeway Mining Corporation built a 270-acre tailings impoundment system (the largest ever constructed) that recycles all cyanide and water in a zero-discharge system. The use of this impoundment system keeps the cyanide and water in a holding compound on the mine property rather than allowing it to flow into the surrounding environment. Efforts are also made to protect birds that fly over the site and wildlife that wander into the mine area by using loud noises to frighten them away.

At the present time, the South Carolina gold mines have only a few years of known reserves left. Unless new ore bodies are found at the Ridgeway Mine, it is expected that all the profitable gold there will have been removed by the year 2001. Scientists have estimated that there are deeper gold reserves left in the Piedmont, and that the Sandhills and other Coastal Plain sediments are perhaps covering up other reachable gold deposits. Continued exploration will determine the future of gold production in South Carolina.

GEOLOGY OF THE GOLDFIELDS

The gold deposits in South Carolina are found in quartz veins; stream gravel or residuum; low-grade, sulfide-bearing quartz replacement deposits; and volcanic-hosted deposits associated with deformation. These are often located in bedrock many feet below the surface.

The quartz veins that contain gold are commonly formed around the boundaries of igneous intrusions, such as granite.

As magma cools, forming the granite pluton deep underground, elements in the melt bond and crystallize to form minerals in an orderly fashion (according to each mineral's crystallization temperature, other minerals in the mix, and the temperature of the progressively cooling magma). Those minerals that have a high melting point crystalize first and include olivine, pyroxene, and calcium-rich plagioclase feldspars. Minerals that crystallize later at cooler temperatures include amphibole, biotite mica, and sodium-rich plagioclase feldspars. The last minerals to crystallize are quartz and rare metals such as gold, silver, copper, and platinum. These last minerals are carried in the hot, watery quartz solution through cracks in the granite and into the nearby surrounding rock. Gold is also emplaced during metamorphic episodes when bedrock is heated, cracked, sheared, and faulted (allowing gold-bearing quartz to travel throughout the damaged rock). When these valuable minerals finally solidify, they form lodes—concentrated pockets or flakes within and near the quartz veins that transported the minerals.

Gold is most easily derived from quartz veins in saprolites as the metal weathers out at or near the surface. The gold particles erode and move downslope. Because gold is 19.3 times as dense as water and six times as heavy as most rock-forming minerals, it readily sinks to the bottom of a creek bed. Placer deposits are often found in streams within the sands and gravels that overlie the bedrock layer.

Gold is also found in the weathered clays of surface saprolite deposits. Because the rocks of the Piedmont have been weathering for so long, there are many old riverbeds and streambeds that lie high above the present streambed levels. These ancient, or "paleo," placers can also be searched out and mined. In South Carolina, these areas are typically 3 to 6 feet thick and overlie saprolites as old as 2 million years (these saprolites are up to 150 feet thick). The gold can lie at the contact or juncture between the saprolite and the old stream deposits.

Most of the gold produced in South Carolina has been found in the Carolina slate and Kings Mountain belts. The Piedmont rocks that contain gold range in age from about 200 million to 550 million years. Some Piedmont granitic plutons are Early Paleozoic, formed when the continental fragment and island arc that formed the Carolina terrain were still far offshore from North America. Other younger plutons, however, were formed as the heat and pressure of later collisions created metamorphic-derived melts, and produced new granites, and hydrothermal solutions that carried and deposited gold into veins. The heat recrystallized many metals, such as magnetite, and introduced sulfur that helped to create the sulfide minerals found in some gold deposits. In South Carolina, these sulfides include pyrite, chalcopyrite, galena, molybdenite, phyrrhotite, sphalerite, and sylvanite.

The rocks in which the gold of the Haile and Ridgeway Mines are found are related. Both mines lie along faulted contacts of the Asbill Pond and Richtex Formations that traverse the Carolina slate belt from southwest to northeast. Gold hunters today follow these same formations to search for additional deposits. The Richtex Formation is made of metamorphosed volcanics (metavolcanics), notably fragmented tuffs, ash flows, and lapilli. These volcanics have been metamorphosed to greenschist facies and include the common products of a relatively low-grade regional metamorphism, which were formed at temperatures approximately 570–930 degrees Fahrenheit. These rocks were deposited both on the land surface and underwater during the Paleozoic when the Carolina slate belt was still offshore. At the Ridgeway site, the gold was emplaced largely in hydrothermal siliceous solutions, and, as a result, the ore bodies consist of up to 90 percent quartz (as a quartzite). The ore bodies are lightly pyritized and they contain abundant white, powdery sericite. Gold particles can rarely be seen in the quartzite, for they are so dispersed throughout that each ton of ore nets only .032 ounces of gold. The south pit of the Ridgeway Mine contains metasediments of the Asbill Pond Formation that are thought to be derived from turbidites—underwater flows of mud, ash, and sands.

The volcanic rocks of the Haile Mine were also metamorphosed into sericite schists interspersed by bands of iron and rutile; however, these rocks are richer in sulfides than those of the Ridgeway Mine. Some geologists believe that the sources of the metal-bearing, sulfide-rich minerals (which are scattered throughout the volcanic rocks) were black smokers—underwater hot springs derived from the Early Paleozoic volcanics of the island arc. Over time, the gold-bearing minerals were leached out and became concentrated below the water table, as a zone of sulfide enrichment deposits.

Other types of rock throughout the Blue Ridge and Piedmont regions also contain gold. These include various granitic rocks, amphibolite, phyllite, quartzite, gneiss, and biotite and chlorite schists. In the Smyrna/Hickory Grove area in York County, gold is found mainly in intrusive quartz diorites (which trend northeast from the area and are found with iron ores and pyrite). In Oconee County, gold is largely found in quartz veins; in Saluda County it has been found in amphibolite and diorite; and in Union County, in phyllite.

Some of the accessory minerals found with South Carolina gold are covellite, enargite, tourmaline, fluorite, pyrophyllite,

Sericite schist near Haile Mine

cuprite, limonite, topaz, calcite, rutile, kyanite, andalusite, bismuth, and malachite.

WHERE TO SEARCH FOR GOLD

Because it takes huge investments and high technology to extract gold from the low-grade sulfide deposits such as those of the Haile and the Ridgeway Mines, the amateur prospector generally searches for free gold that does not need refining to be seen. This native gold comes from the surface and near-surface oxidized zones, where it has become concentrated in streambed sediments and saprolites as a result of weathering and erosion.

For those who only wish to explore and pan on a small scale, there are many streams and rivers in the Piedmont that can still provide "color." Some of the streams in the upstate that have produced gold in the past include the Middle Tyger River in Greenville County, Mountain Creek in Spartanburg County, Broad River in Cherokee County, Sleepy Creek in Edgefield County, and Cedar Creek in Fairfield County.

Would-be prospectors should always ask for permission before going on private land and they should not try to pan or collect minerals or rocks in state parks! National forests, however, will allow moderate gold panning without special permits. No fee, special permission, or permit is required as long as the specimens are for personal, noncommercial use and they have no archeological value, no mechanical equipment or blasting is used, no surface or stream disturbance results, and the collection does not conflict with existing mineral permits or leases. The National Forest Service writes in its brochure, "Rockhounding on National Forest Land in National Forests in South Carolina":

> Included with rockhounding is panning for gold in the beds of many streams crossing national forest land. . . . The Southern Region allows gold panning in the beds of most streams crossing national forest lands. No fee, special permission or permit is required as long as

only shovel and pan are employed and no significant stream disturbance results, but one should check with the district ranger first. On national forest land, where the mineral rights are leased, panners should obtain written permission from the mineral owners prior to beginning.

Lists and maps of old gold diggings located on private lands can be purchased (see resource list). With a bit of effort and a little luck, a persistent prospector might yet discover another South Carolina gold mine!

Harpers New Monthly Magazine, August 1857: "Bill Jenkins" mining for gold

How to Pan for Gold

1. Using a shovel, dig some sand and gravel from the area just on top of bedrock from the bottom of an old elevated streambed or from a running streambed.
2. Put the sediments into a gold pan (available from rock shops or commercial gold mines).
3. Submerge the pan and its sediments in water, break up any clays, and toss out pebbles that are marble-sized or larger.
4. Settle any gold that might be present by shaking the pan side to side about a dozen times under water. Rinse.
5. Add water, tilt the pan forward or toward you. Repeat several times, each time pouring off water and sediment.
6. Wash the contents of the pan until only a few small rock fragments remain with white and black sands. The gold will be under the black sands.
7. Carefully tip the pan down, swirl the black sands gently with a small amount of water. This water should wash toward you over material on the far side of the pan, exposing the gold.
8. Carry a glass vial to hold your gold. Fill it with water and deposit into it your flakes and nuggets!

Peachtree Rock, Lexington County

CHAPTER 5

THE GEOLOGY OF THE COASTAL PLAIN

The Coastal Plain region comprises about two-thirds of South Carolina's land area. It lies between the Sandhills and the coast, and is part of the larger Atlantic Coastal Plain that runs from New Jersey to Florida and west to Texas. This well-ordered wedge of sediments begins at the Sandhills—about 640 feet above sea level in Saluda County and 725 feet above sea level in Chesterfield County—and runs south and southeast to the Atlantic Ocean. The Coastal Plain in South Carolina includes a variety of landforms from rolling, sandy hills to plateaus, plains, river valleys, swamps, Carolina bays, estuaries, tidal creeks, marshes, and barrier and sea islands.

The Coastal Plain is made up of Mesozoic and Cenozoic sediments and sedimentary rocks that overlap the crystalline rocks of the Piedmont. These sediments are thinner toward the Fall Zone and thicker toward the coast. They are both marine and terrigenous sediments—deposits made as the sea encroached time and again across the land and as rivers eroded the Blue Ridge Mountains to the northwest and carried their sediments seaward. The Coastal Plain is divided into three sections according to elevation—upper, middle, and lower.

THE UPPER COASTAL PLAIN

The geology, and to a large measure the geomorphology of the upper Coastal Plain, has been shaped by the warping of the land under South Carolina. Beginning in the Paleocene, the Coastal Plain sediments were tilted up toward eastern North Carolina to form the Cape Fear Arch and were tilted down toward Georgia to form the South Georgia Embayment. Over millions of years in this tilted position, erosion became pronounced in South Carolina's eastern region, and deposition became pronounced toward the southwest. The sediment depth under Little River at the North Carolina–South Carolina state line is 400 feet, while under Hilton Head Island, near the Georgia border, it is 4,300 feet.

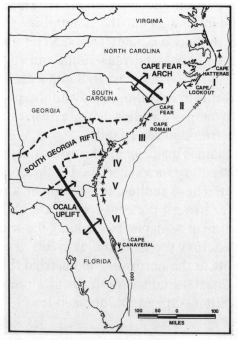

Map showing Cape Fear Arch (courtesy: Dr. Miles O. Hayes)

The geology and geomorphology of the upper Coastal Plain were also affected by several episodes of encroachment and abandonment by the sea. During these episodes, its stratigraphy was developed by repeated cycles of deposition and erosion that, over the past 90 million years, produced such features as barrier islands and back-barrier marsh formations, river valleys, deltas, lake deposits, shallow sea floor terrain, beaches, dunes, and escarpments. To decipher the complex assortment of geomorphic structures is difficult because the repeated transgressions and regressions of the sea erased and overlapped those features created by the normal, ongoing continental erosional and depositional processes of the rivers and streams.

River deposits, S.C. Highway 378, east of Horrell Hill, Richland County

"Citronelle Gravels" of an ancient river-bed, Miocene in age, found in cliff above Horrell Hill in Richland County

The upper Coastal Plain, beginning at the Sandhills, forms a hilly terrain that is dissected by large rivers such as the Congaree and Wateree. It ends at the Orangeburg Scarp—a wave-cut ridge that traverses the state diagonally at elevations that range from 180 to 215 feet above sea level.

The sediments of the upper Coastal Plain represent the work of millions of years of erosion and deposition by rivers. As discussed in chapter 1, the Sandhills were formed in part by wind-blown sands provided by these rivers as they brought sediments down from the higher elevations of the Blue Ridge and Piedmont. The ages of the Sandhills sediments range from Cretaceous to Miocene. The oldest sediments on the upper Coastal

Middendorf Formation of kaolinite and sandy clays, north of Patrick on U.S. Highway 1, Chesterfield County

Sugarloaf Mountain sandstone remnant

Plain are found in the Cretaceous Middendorf Formation. This formation is composed of kaolinitic clays and clayey, quartzose sands that were deposited approximately 100 million years ago. This clay layer can be seen along road cuts atop Piedmont bedrock (for example, along U.S. Interstate 20 in the Columbia area). Excellent outcrops of the Middendorf Formation are found around Sugarloaf Mountain in Chesterfield County. Part of the Sand Hill State Forest, Sugarloaf Mountain rises 100 feet above the road and is 513 feet above sea level. It is thought to be an erosional remnant of a Miocene oceanic inlet or embayment in the Orangeburg Scarp. Composed of clays, clayey sands, and sands, it is capped by iron-cemented sandstone. Bright white kaolinic clay can be seen at the base of the formation along U.S. Highway 1.

The Pinehurst Formation forms the top layer of the Sandhills. It is made of ten-million-year-old Pliocene sand deposits. The evidence of crossbedding within the dunes; the lack of marine

U.S. Silica Mine near Edmund, in Lexington County, on S.C. Highway 302

fossils; and the scarcity of mica, clay, and silt point to deposition by the southwest prevailing winds. As a result, extremely pure silica sands have been produced and are mined in areas including Lexington County. At sites such as the U.S. Silica deposits, the sands are so vast in area that geologists believe their exploitation can continue for the next 100 years.

The upper Coastal Plain is further divided into the Aiken Plateau, the Richland Red Hills, and the High Hills of Santee. All of these uplands are remnants of land left after the transgressing seas of the Pliocene retreated. The Aiken Plateau, which lies between the Savannah River Valley and the Congaree River Valley, is a raised plain that slopes gently seaward at elevations ranging from 600 feet in the northwest to 300 feet in the southeast. It includes parts of Aiken, Edgefield, Saluda, and Calhoun Counties. The Aiken Plateau provides the most complete lithologic history of those sediments deposited between the Cretaceous

Contact between Late Cretaceous and Paleocene sedimentary layers, S.C. Highway 601 on east side of Fort Jackson, Richland County

Period and the Pliocene Epoch. The various formations that make up this plateau are composed of sediments that were deposited in upper deltaic river environments, on shallow ocean shelves, and in offshore deep waters. The kaolinite clay beds of Aiken County were formed during the Late Cretaceous and the

Eocene in coastal delta settings. Evidence of this environment can be seen in marine diatoms and tree and shrub leaves contained in the clay.

One of the most interesting sites on the Aiken Plateau is Peachtree Rock, on the eastern slope of the plateau in Lexington County. This small 305-acre preserve was created in 1980 by the South Carolina Nature Conservancy. It provides one of the best views found in the state of an ancient nearshore Eocene marine environment. Peachtree Rock is an eroded remnant of Middle Eocene multicolored marine sands and clay lenses that has been protected from destruction by its weather-resistant silicified upper layer. The area between Peachtree Rock and the road is made up of upper Eocene sands, Miocene river deposits, and Pliocene/Pleistocene sand dunes. Nearby are several other sandstone exposures that contain marine fossils, mainly in the form of shell hash, and extensive shrimp burrows. Fossil collecting is not allowed in the preserve.

The Richland Red Hills are found across the Congaree River valley from the Aiken Plateau. East of these hills, just across the Wateree River valley, lie the High Hills of Santee that span portions of Sumter and Lee Counties. Both of these highlands are highly eroded remnants of the Aiken Plateau and contain upper delta plain and marine sediments. Both areas can be seen by driving east from Columbia on U.S. Highway 378 toward Sumter.

The upper Coastal Plain ends at the Orangeburg Scarp. The escarpment overlooks the city of Orangeburg as a high ridge. It is a dominant feature of the southeastern Coastal Plain. The escarpment continues northeastward through Sumter, Hartsville, and Bennettsville, to North Carolina, and southwestward through Allendale to Georgia and Florida. Because the transgressing Pliocene sea formed embayments as it moved up the river valleys (including the Congaree and the Wateree), the escarpment is also found under the University of South Carolina and the state capitol building in Columbia—42 miles north of Orangeburg.

THE MIDDLE AND LOWER
COASTAL PLAIN

The surfaces of the middle and lower Coastal Plain are sculpted by six step-line escarpments and seven terraces that represent seven cycles of ocean transgression and regression— two in the Pliocene, four in the Pleistocene, and at least one in the Holocene. Terraces are temporary ocean floors that formed as the sea level rose and fell over time. Each terrace is headed by its own escarpment because escarpments represent the landward limit of the sea at the time each terrace formed. In order of position, beginning at the base of the Orangeburg Scarp and traveling toward the coast, the terraces are the Coharie, Sunderland, Okefenokee, Wicomico, Penholoway and Talbot, Pamlico, and Princess Anne. The escarpments are the Orangeburg, Parler, Surry, Dorchester, Summerville, and Bethera. The escarpment forming at present sea level is the Cainhoy.

The middle Coastal Plain begins at the Coharie Terrace and continues to the Surry Scarp—the 130-foot boundary between the Okefenokee and Wicomico Terraces. The Surry Scarp was formed during the most recent Pleistocene submergence of the coast, 85,000 years ago. It forms a wave-cut ridge in places, but it also represents a barrier island shoreline and a beach and the valley wall of a river.

The lower Coastal Plain begins at the Wicomico Terrace and continues to the coast. The rocks and sediments exposed in the middle and lower Coastal Plain range in age from Cretaceous to Holocene. The oldest Cretaceous rocks are exposed along the Pee Dee River and the youngest Holocene sediments are exposed in river sediments and the beach sands along the coast. For a detailed description of the geology and geomorphology of the Coastal Plain, refer to Horton and Zullo, *The Geology of the Carolinas*. The geological map of South Carolina (see insert in this chapter) shows the general positions of the major rock and sediment formations of the Coastal Plain.

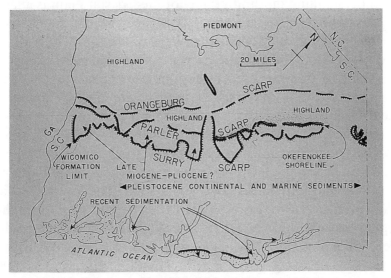

Ancient shorelines of the middle and lower Coastal Plain, after D. Colquhoun (courtesy: Dr. Miles O. Hayes)

Mapping the various rock formations of the Coastal Plain is ongoing, with the use of surface mapping techniques and core hole readings. Geologists continue to identify the types of exposed sediments and rocks and the geologic structures that contain them to decipher the complex history of the Coastal Plain. They are correlated in the following manner:

Sediment Type	Ancient Depositional Topography
conglomerates, sand, silt	floodplain
sand, clay	barrier island
quartz sand	beach, nearshore continental shelf, windblown sand
quartz sand, silt, clay, marl	offshore continental shelf
limestone	offshore continental shelf

Understanding the stratigraphy of the Coastal Plain is a task of (1) sorting out the nature of the erosional and the depositional environments found in any given section of the plain, and (2) relating them to the transgressions and regressions of the sea and the more recent actions of streams, rivers, and marshes. Rapid drops in the contour of the land are found in the Coastal Plain and have resulted from wave scour. Many hills and depressions are remnants of barrier islands and marshes; gravels found atop hills are ancient riverbeds; and level plains that stretch for miles are the remains of sandy ocean bottoms that long ago formed the continental shelf. The lower Coastal Plain offers some of the most interesting geomorphology in the region because the original shapes of the land, although modified, are still visible. The dates of origin of the lower Coastal Plain surfaces range from Pleistocene (1.8 million to 10,000 years ago) to Holocene (10,000 to 6,000 years ago).

As the seas transgressed over the land, deposits of limestone were made on the continental shelf. This shelf today is exposed as the Santee Limestone Formation, found in much of the middle Coastal Plain. One of the most interesting areas to visit within this formation is the limestone karst region in and around Santee State Park—an area of sinkholes, small caves, and underground streams. This characteristic karst topography formed as acidic groundwater percolated through the porous rock and chemically reacted with the limestone to dissolve it. There are some large

An example of karst topography: sinkhole filled with water found in Santee State Park

karst topography: small sinkhole in the woods of Santee State Park

karst topography: underground river exits mouth of small cave found in Santee State Park

karst topography: looking down a chimney to an underground river in Santee State Park

sinkholes in the Park that formed as the roofs of the caves lost their support from below and collapsed. While not as large as karst features in Florida, or in other parts of the country, some of the Santee State Park sinkholes are as big as a house. They clearly illustrate the weathering and erosion of limestone by acidic water.

The transgressions and regressions of the seas have eroded much of the sediments of the middle and lower Coastal Plain in South Carolina. Beginning in the Oligocene, the coastline was located over 50 miles out on today's continental shelf. The land was at that time eroding, not depositing, sediment—except in river valleys. Sea levels dropped around the world during the Oligocene as extensive ice caps formed in areas such as Antarctica. The ice caps trapped water from the oceans causing the sea levels to fall over 100 feet. The fall of sea levels and the increased stream and river erosion resulted in huge unconformities of millions of years between sediment layers on the Coastal Plain in South Carolina. By drilling holes in various parts of the Coastal Plain, geologists have brought samples of deep-lying sediments and fossils to the surface. Using these samples, they have created models of and dated the formations buried beneath the surface. They have also determined the extent of each formation and the nature of the incomplete or missing layers.

UNDERGROUND WATER IN THE COASTAL PLAIN

The abundant underground water of the Coastal Plain is one of the state's most valuable resources and it is of great interest to environmental geologists. Most of the water is stored in permeable rock, called aquifers, made up of sediments such as sand and limestone. As they understand the various strata and the lithology of the Coastal Plain, geologists can map areas of stored underground water. Water that feeds the aquifers flows in from a catchment (or intake area) of rivers and streams above the aquifer. The water is held in place by aquicludes—which are

EXPLANATION OF DRILL–HOLE LOGS

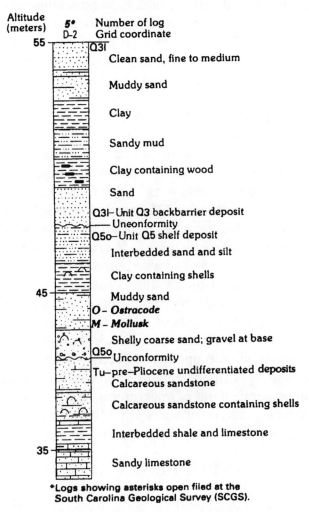

Diagram: U.S. Geological Survey drill-hole logs (courtesy: Department of the Interior, U.S.Geological Survey)

layers of dense, impermeable clays that will not allow water to pass quickly. Since the water flows downpressure, it can move uphill as well as downhill from a site of pollution. Mapping of aquifers is vitally important in order to preserve them because South Carolina has many sources of toxic pollutants that endanger these water supplies: nuclear waste, industrial and urban sewage, agricultural runoff, and leaking underground storage tanks. Because it is of such vital concern to maintain the purity of the underground water reserves, the aquifers in the state are monitored by the companies that store or discharge toxic waste. These companies are, in turn, monitored closely by the state and federal governments.

In recent years, the Savannah River Site became a case in point. The site is built above an aquiclude of clay that was believed would prevent the ground water from carrying radioactivity to the underlying aquifers. However, citizens in Georgia, across the river from the site, were surprised to find traces of radioactive tritium gas in their deep 225-foot wells in 1991. Scientists are still investigating the maze of aquifers and aquicludes, as well as possible surface paths, to determine how the tritium traveled so far and reached so unexpectedly deep.

Overuse is another danger to the Coastal Plain water supply. Because agriculture, industry, and urban centers all use and dispose of tremendous amounts of water, the South Carolina Water Resources Commission must monitor the use of the state's water supply. Without proper and effective management of the available resources, both humans and wildlife would suffer. Water can be pumped out of underground aquifers quickly, but it takes nature hundreds of years to replenish them. Once polluted, these aquifers are very difficult, if not impossible, to clean.

COASTAL PLAIN RIVERS

South Carolina has 9,900 miles of rivers, 525,000 acres of lakes, and 2,155 miles of estuaries. Many of the rivers that flow across the Coastal Plain—such as the Wateree, Congaree, and Savannah—have been dammed upstream to create large lakes. These rivers deposit much of their sediment load in the lakes behind the dams and carry mostly the finer sediments to the sea. As they wind their way toward the coast, the rivers slow down and leave behind the mud and sands that create the spectacular estuarine environments for which South Carolina, and much of the South Atlantic coast, is famous.

Most of the larger rivers of South Carolina belong to systems that originate in North Carolina. In contrast, the rivers that begin in the Coastal Plain are of a different type—they are the black rivers, which include the Edisto, Combahee, Salkehatchie, Black, and Coosawhatchie. Because these rivers are not moving from very high elevations to low elevations, and because they have short drainage basins, they are free of heavy sediment loads. Their waters appear dark because their beds are loaded with humus—organic matter that darkens its waters with tannins, much like the effect of tea in a cup. Early plantation owners used these black rivers to build the rice beds that created much of the wealth of colonial South Carolina. These colonial fields can still be seen near many of the coastal river estuaries such as the Waccamaw, Ashley, Cooper, Edisto, and Combahee.

The longest black river in North America, the Edisto, begins in Edgefield County and runs southeastward about 150 miles. Adventurers may now easily enjoy the placid beauty of the Edisto River by canoe or kayak. A trail has recently been established on the river from Whetstone Crossroads on U.S. Highway 21, through Colleton State Park southeast to Galivants Ferry State Park.

Diagram: Edisto River Kayak Trail (courtesy: *The State*, Columbia, South Carolina)

Because of the extensive low elevation and proximity to the sea, the Coastal Plain contains some of the most spectacular wetlands in all of North America. South Carolina has excellent examples of the great swamps that stretched all along the Coastal Plain of the eastern United States as far west as eastern Texas before the late nineteenth century. Two swamp preserves that remain intact and undeveloped, and highly accessible to the public, are Four Holes Swamp (near Harleyville in Dorchester

Alligator in Coastal Plain swamp (courtesy: South Carolina Department of Parks, Recreation, and Tourism, Tourism Division)

County) and Congaree Swamp (outside of Columbia, in lower Richland County).

Four Holes Swamp meanders 62 miles toward the Edisto River from Calhoun County. The Francis Beidler Forest found within the Four Holes Swamp is a 5,820-acre, pristine wetland that contains the largest remaining virgin stand of bald cypress and tupelo gum trees in the world. Some of these cypress trees are over 1,000 years old. The forest is home to many protected birds and other endangered animals and rare plant life. The black water swamp is fed by runoff and by springs, so its water levels fluctuate greatly with seasonal rainfall and drought.

The Congaree Swamp, on the Congaree River floodplain, contains the oldest significant stand of old growth river bottom hardwood forest in the United States. Over ninety species of trees grow in this 22,000-acre national monument—including loblolly pine, bald cypress, water tupelo, green ash, American elm, sweet gum, and many varieties of oak. The swamp floods an average of ten times a year, most often in February. These floods deposit rich silts and clays creating the soils that nourish its bottomland forests. The rich floodplain environment also provides habitat for many species of animals—including bald eagles, alligators, birds, and bobcats.

The Congaree River is formed as the Broad and Saluda Rivers join at Columbia. The Congaree then flows 60 miles across a floodplain, eroding mainly on its eastern bank, to merge with the Wateree and form the Santee River. Because the Congaree River floodplain decreases in elevation only 10 feet in more than 13 miles, the river, rather than downcutting its riverbed, is forced instead to wander over it. As it moves slowly over the floodplain it forms great curves, called meanders, which over time are abandoned as the river changes its course due to sediment

Cypress knees in Congaree Swamp

accumulation and floods. Some of these abandoned meanders then form oxbow lakes. The Congaree Swamp contains two excellent examples of oxbow lakes that are easily accessible to visitors—Weston Lake and Wise Lake. It is estimated that Weston Lake became an oxbow lake only about 10,000 years ago.

Another landmark on the Coastal Plain is the ACE Basin, located 45 miles south of Charleston. It is a near-pristine preserve made up of the lower few miles of the Ashepoo, Combahee, and Edisto Rivers and their estuaries. The ACE Basin Preserve comprises 350,000 acres of diverse habitats including upland hardwood forests, forested bottomland, 91,000 acres of marshes, former rice fields, and estuaries. It is home to over fifty distinct natural communities and it contains over five hundred plant and animal species—including seventeen endangered species such as the southern bald eagle, the loggerhead turtle, and the wood stork. Also found in the area are coyotes, alligators, bobcats, and mink.

The ACE Basin was established in 1992 as a protected area due largely to the efforts of many South Carolina landowners, the South Carolina Nature Conservancy, Ducks Unlimited, the South Carolina Department of Natural Resources, and the United States Fish and Wildlife Service. A total of 9,695 acres of the basin have been set aside as South Carolina's fifth national wildlife refuge, and 11,942 acres have become the National Estuarine Research Reserve. There are presently 16,661 acres of protected sites, public and private, within the basin. Its preservation

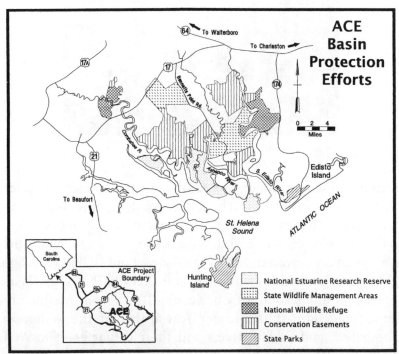

ACE Basin protection efforts (credit: South Carolina Department of Natural Resources Newsletter, Summer 1994)

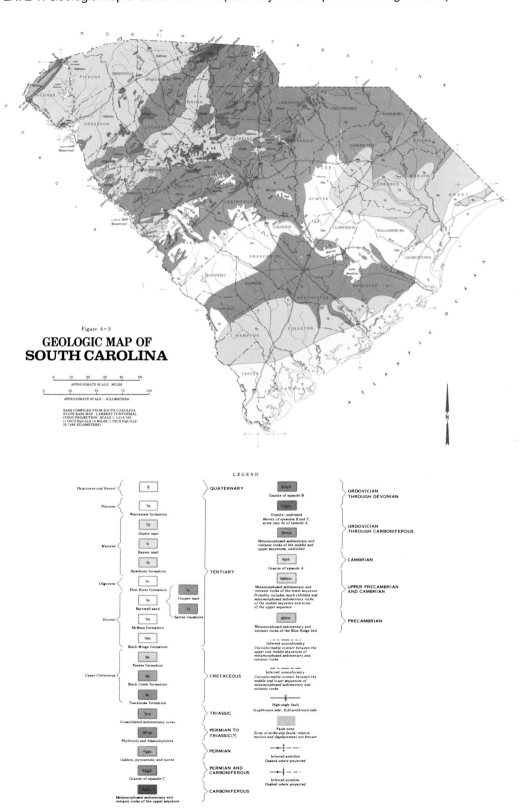

Figure A - 5

GEOLOGIC MAP OF SOUTH CAROLINA

PLATE 2. Geologic Belts in South Carolina (courtesy: U.S. Department of Agriculture)

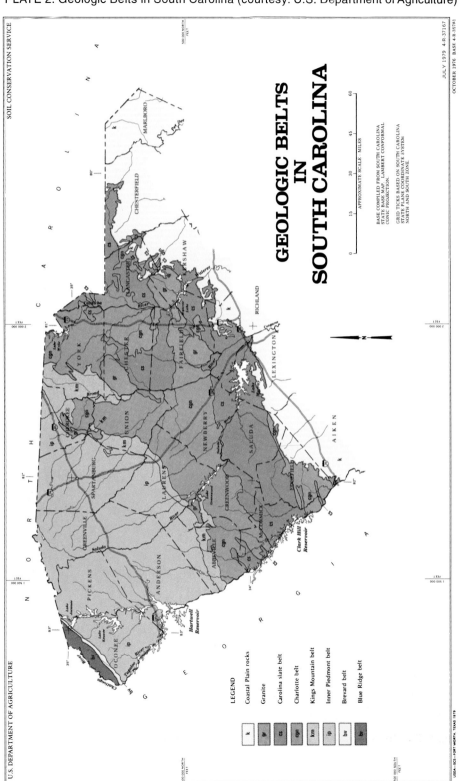

PLATE 3. Blue Ridge Province, gneiss near Whitewater Falls

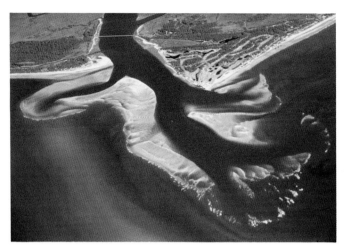

PLATE 4. Ebb tidal delta at Fripp Island. Note extensive channel margin bar on the right (courtesy: Dr. Miles O. Hayes)

PLATE 5. Carolina Bay (courtesy: South Carolina Land Resources Commisions)

The ACE Basin (courtesy: Dr. Miles O. Hayes)

safeguards one of the most remarkable and unspoiled ecosystems to be found anywhere in North America. It is open to the public for recreation, education, and research.

ROCK AND MINERAL RESOURCES
OF THE COASTAL PLAIN

The Coastal Plain provides an abundance of resources for South Carolina and it generates a great deal of income for the state. Foremost among products mined in the Coastal Plain are sand, gravel, limestone, crushed stone, kaolin, and other clays.

Sand mining accounts for 300 of South Carolina's 486 mining operations, and it is by far the largest component of its mining industry. Within the middle and lower portions of the Coastal Plain, sand and gravel are mined in many areas such as Cheraw, Bennettsville, Blenheim, and Johnsonville near the Pee Dee River; areas along the Wateree River; and the Savannah River near Augusta, Georgia, where floodplain deposits are plentiful. The sands are used as building construction material and the gravels are used mainly in road construction. Fine silica sands are mined largely from dune deposits in the upper Coastal Plain. The mines in Lexington County are among America's largest producers of finely milled silica used in fiberglass, computer microchips, and paint filler.

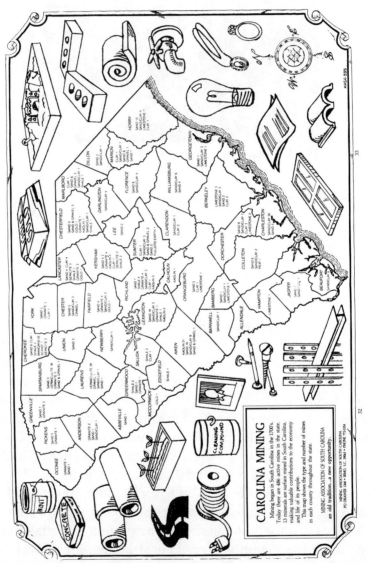

Mining resources map (courtesy: Mining Association of South Carolina)

Limestone has been produced in South Carolina since 1820 and it is mined today at many sites in Georgetown, Horry, Clarendon, Orangeburg, Dorchester, and Berkeley Counties. Because of the various geological conditions under which limestone developed in South Carolina, it is found in many forms. Horry and Clarendon Counties mine largely coquina—an organic limestone formed from a mixture of shells and lime used largely for roadbeds. Most of the limestone found in Orangeburg and Dorchester Counties is a marl—a mixture of lime and clay. At the Giant Portland Cement plant near Harleyville, marls are mixed with additional clay and made into Portland cement. Today, South Carolina is the leading producer in the Southeast of this very important construction material. With 1993 sales of $100 million, limestone products are the most valuable mineral commodity in South Carolina. Much of the limestone mined in Berkeley County is used as crushed stone. Crushed stone, with sales of $98 million is second, and gold, at $73 million, is third.

South Carolina ranks second in the nation in the production of kaolin clays. The state's rich kaolin beds furnish about 12 percent of the total kaolin clay mined in the United States. South Carolina has thirty active mines, but most kaolin is mined in the upper Coastal Plain in sites near the towns of Aiken, Bath, and

Aiken clay beds

Langley. This valuable mineral has been used since before the Revolutionary War, when it was shipped to England to be used in Wedgewood pottery. Large users of clay include the brick and rubber industries, as well as the paper industry that uses kaolin as a coating to produce glossy paper. Clay is also used in medicines, cosmetics, and ceramics.

Related to kaolin, fullers earth is a general term for highly absorbent clays. Since fullers earth can be made even more absorbent by heat, it is used in the purification of crude oil, the manufacture of floor-sweeping compounds, and as a filler for rubber and plastics; but, it is perhaps best known as the main

Fullers earth at Pinewood underlies red clay and can be seen in the liner wall (courtesy: Laidlaw Environmental Services of South Carolina, Inc.)

ingredient in kitty litter. Near Pinewood in Sumter County, the fullers earth is being used as a receptacle liner for chemical and other toxic waste.

Phosphate was once an important mineral commodity that was mined for many years in the Charleston and Beaufort areas for use as agricultural fertilizer. From 1867 to the 1930s, phosphates were dredged from thin beds only 3 to 36 inches thick—mainly from the Wando, Cooper, Ashley, Edisto, and Coosaw River Basins. These phosphates formed in Oligocene, Miocene, and Pliocene seas as the tides and currents of the Gulf Trough deposited organic oozes on the shallow seabed. These oozes contained calcium phosphate, which combined with lime from the underlying marl to form phosphate of lime. Carbonic acid was released in the process, which dissolved portions of the uppermost few inches of the underlying limestones and leached into the subsurface forming a thin pan of phosphates. Although it is estimated that millions of tons of phosphates can be found from offshore South Carolina to the Blake Plateau, the phosphate industry is no longer active in the state.

Peat is a minor industry. It is produced in Colleton County from river and bog deposits. This organic material is used as a soil conditioner, and in wastewater treatment as a very effective filter for coliform bacteria, heavy metals, and phosphates.

Rare earth elements—such as gallium, beryllium, and titanium—are found in minerals such as rutile, zircon, and ilmenite in placer deposits in stream valleys in the upper Coastal Plain. These minerals, eroded from the granitic rocks of the Piedmont and Blue Ridge, were dredged for a time in the 1950s. They are not economically recoverable at this time; however, they are present in large enough quantity to justify the designation of the upper Coastal Plain as a "rare earth province."

The Black Creek Group in the middle Coastal Plain has been explored for the presence of uranium. Deposits have been found within its clays and sandy Late Cretaceous sediments in Dillon County. These uranium deposits may become economical and environmentally safe to mine in the future, but there are no known plans to mine them at this time. Other uranium sources include the phosphates that are found along the coast and offshore on the continental shelf.

It has been estimated that 7 percent of America's undiscovered, recoverable petroleum reserves and 4 percent of its natural gas reserves are located on the Atlantic continental shelf and continental slope. Many geologists feel that about one-fourth of these petroleum reserves and one-third of the gas reserves are located in the South Atlantic region. There is a possibility, then, of future exploitation of the reserves off the South Carolina coast.

OTHER RESOURCES
OF THE COASTAL PLAIN

South Carolina agriculture is dependent upon its soils, although most are below average in richness. The soils of the upper and middle Coastal Plain are well-drained, mostly loamy sand with sandy clay subsoils. While poor in organic content, these soils are the best in the state and sustain South Carolina's most productive farmland. Toward the coast, running in an irregular belt 30 to 70 miles wide, the soils are moderately to poorly drained loamy sand to sandy loam, with clay subsoils. Soil fertility is low to medium and acidity levels vary from strong to very strong; therefore, much of this land must be left to forest, swamp, and marsh. With artificial fertilization, however, many acres of sandy Coastal Plain soils produce soybeans, tobacco, cotton, corn, wheat, and vegetables as truck crops. Much of the

Coastal Plain tobacco farming, Pee Dee region (courtesy: South Carolina Department of Parks, Recreation, and Tourism , Tourism Division)

lower Coastal Plain has been planted in fast-growing longleaf pines that fuel South Carolina's pulp industry. The soils of the Coastal Plain used for farming are an invaluable state resource.

The greatest attraction of the Coastal Plain might very well be its natural beauty, given that tourism is South Carolina's premier industry. Every year, millions of people flock to the coastal islands, beaches, and waterways for recreation. In 1780, South Carolina had 6.4 million acres of wetlands; but, in 1980 it had only 4.6 million acres left. Happily, 40 percent of the coastline is now held in trust—either as wildlife preserves or as public recreation areas. Many citizens, nonprofit groups, and government agencies are working to protect the future of the state's rare wetlands.

The Coastal Plain contains many of the most interesting geological sites in the state. It offers wetlands, islands, rivers, rocks, and minerals for the lay visitor to study and enjoy, as well as surface and subsurface structural questions that still puzzle geologists. New species of fossils continue to be discovered in the sediments of the Coastal Plain. Seismologists still study the structure and movements of the land on and offshore. And the search to find ways of safeguarding the wetlands and aquifers moves forward. The Coastal Plain will remain a challenging laboratory for the professional and the student, and a place of beauty for the visitor.

Interesting Places to Visit on the Coastal Plain

There are many state parks on the Coastal Plain. Refer to the list provided by the South Carolina Department of Parks, Recreation, and Tourism for a more complete listing.

Upper Coastal Plain

Aiken area—kaolin mines along U.S. Highway 1 west of Aiken—clay beds along roadsides.

Fall Zone rapids—Columbia—rapids visible from the Riverbanks Zoo and at various river landing sites on the Saluda, Broad, and Congaree Rivers.

South Carolina State Museum—Gervais Street in Columbia—wide variety of science displays; South Carolina fossil display, including dinosaur remains.

McKissick Museum—on the Horseshoe, Sumter Street, University of South Carolina in Columbia—rock, mineral, and fossil displays.

Horrell Hill to Congaree Swamp—To travel from the upper to lower Coastal Plain on the south bank of the Congaree River Valley, follow S.C. Highway 769 south from U.S. Highway 378 through the town of Congaree to S.C. Highway 48 and Old Bluff Road to the Congaree Swamp National Monument, in Richland County.

Peachtree Rock Preserve—Follow S.C. Highway 302 south from Columbia to S.C. Highway 6 past the town of Edmund, then left near Bethel Methodist church onto dirt road; preserve is on the left; in Lexington County—Eocene sandstone formation.

Sand Hill State Forest—U.S. Highway 1, 4.3 miles west of Patrick in Chesterfield County—location of Sugarloaf Mountain, a late Cretaceous erosional remnant.

Middle and Lower Coastal Plain

ACE Basin Preserve—south of U.S. Highway 17,beginning in the communities of Green Pond and Ashepoo, in Colleton County—pristine blackwater rivers and drowned coast topography.

Barrier Islands—see chapter 6.

Carolina Bays—see chapter 7.

Charleston Museum—Meeting Street—fossils, rocks, and minerals.

Congaree River—swamps, meanders, and sand deposits below Columbia; boat rentals available in Columbia.

Congaree Swamp National Monument—southeast of Columbia off S.C. Highway 48 in Richland County—extensive swamp forests, and lake and river environments.

Crowburg Basin—on S.C. Highway 207 west of Pageland in Chesterfield County—Triassic red beds.

Edisto River—Colleton State Park on S.C. Highway 61 north of Walterboro in Dorchester County—kayak and canoe trail.

Fossil collecting—see chapter 8.

Four Holes Swamp: Francis Beidler Forest—off U.S. Interstate 26 at Harleyville, then S.C. Highway 28 in Dorchester County—oldest virgin stand of cypress trees in North America.

Francis Marion National Forest—U.S. Highway 17 north of Charleston in Berkeley County.

Healing (Artesian) Springs—north of Blackville on S.C.Highway 3 in Barnwell County.

Santee State Park—on shores of Lake Marion off S.C. Highway 6 in Orangeburg County—karst topography (limestone sinks).

Savannah River Bluffs Heritage Preserve—off S.C. Highway 230 near North Augusta in Aiken County—river shoals and bluffs.

Stateburg, High Hills of Santee—U.S. Highway 378 at S.C. Highway 261 in Sumter County.

Waccamaw River—S.C. Highway 544 west of Myrtle Beach in Horry County—rice plantations along the shore; boat trips from Socastee.

Morris Island lighthouse lies 1,600 feet offshore from the north end of Folly Island

CHAPTER 6

THE BARRIER AND SEA ISLANDS

The barrier and sea islands that lie along South Carolina's coast are part of one of the longest coastal island chains in the world—stretching from Maine to Mexico for 10,000 miles. These islands are major attractions of the Carolina coast, and they are famous as sites for much of South Carolina's early history. Events—including the founding of the Carolina colony at Charleston, campaigns fought by Americans and British during the Revolutionary War, and the shots fired at Fort Sumter that began the Civil War—occurred on or near the barrier and sea islands and their marshes and inlets. Today these islands are important to South Carolina in a less dramatic, but no less important, way—they provide the basis of much economic development from tourism, recreation, fishing, and real estate.

South Carolina's 187-mile coast has thirty-five barrier and sea islands, totaling 144,150 acres. Of the southeastern states, only Florida has more offshore islands. The South Carolina chain begins with Waites Island at Little River Inlet in Horry County, and runs southeast to the Georgia border, ending at Jones Island at the Savannah River in Jasper County. Most of the barrier islands lay along the southern coast within a large recessed shore, or bight, called the Georgia Embayment that runs from Cape Romain to northern Florida.

Myrtle Beach tourists enjoy the beach (courtesy: Tim Kana)

THE ORIGINS OF BARRIER ISLANDS

The birth and evolution of barrier islands are dependent upon several factors—large-scale tectonism, local geologic history, sediment supply, relative sea levels, tidal and wave energy flux, and inlet switching created by storms. Seventy-six percent of barrier chains are located on passive (or trailing) continental margins, and most are within tidal range areas of less than 13 feet. South Carolina lies on a passive continental margin; the mean tidal range and wave energy along its coast, especially within the Georgia Embayment, is moderate. The tidal range is lowest along the Grand Strand and increases southward toward the Georgia border. This combination of moderate tides and low wave action within the Georgia Embayment allows many barriers islands to exist.

The present Holocene barrier islands along the South Carolina coast began to form as the rapid postglacial sea level rise began to slow. During the last glacial period (the Wisconsin), sea levels were far lower than today, for water that had previously been in oceans was trapped as ice. Rivers eroded a great deal of sediment from the land and deposited it on what is today the continental shelf. When the Ice Age ended about 15,000 years ago, the glacial ice began to melt and world sea levels then rose for thousands of years. About 6,000 years ago the sea level rise began to slow, and, as it did, the erosional action of waves deposited sand as long sandbars. As more sand was added over time, the sandbars rose above water to become barrier beaches and then larger and higher barrier islands. Pleistocene formations lie at the base of most of the barrier islands.

THE SEA ISLANDS

South Carolina, Georgia, and Florida also have sea islands. These are islands formed out of mainland sediments that were surrounded by water as sea levels rose. The barrier islands are attached to these pieces of isolated mainland on the seaward side. The core of the sea island is made up of thick continental sediments and soils that are covered with dense forest—often oak, pine, gum, and cypress trees. The soil on the sea islands is rich enough to support farming, as seen 150 years ago in South Carolina with the greatly prized sea island cotton. Today, small farms produce vegetables and fruit on parts of the islands, and also harvest some pines and hardwoods. Developers have found that, instead of agriculture, the valuable and scarce acreage of the sea islands is best used as real estate for vacation and retirement homes and recreation. As a consequence, the struggle to develop as much of the islands as possible is on, and the battles between developers and conservationists are protracted and intense. Development of islands such as Hilton Head is a case in point. Like other areas along the coast, recent depletion of the water table by the heavily concentrated population has led to saltwater migration into aquifers. This loss of potable water threatens future development on the islands, as well as those areas on the mainland that use the same underground water sources. The issues such as water and land use demand creative problem solving and will undoubtedly continue far into the next century.

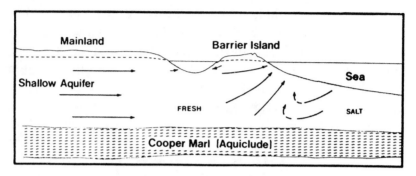

Diagram: Schematic cross section through the shallow aquifer for the eastern portion of the Charleston area showing the circulation of seawater and the general position of the zone of diffusion between fresh water and salt water (courtesy: Tim Kana)

The present islands are only the most recent that have followed South Carolina's mobile shoreline. Evidence of a long history of growth and movement of barrier islands is clear—for their long, thin shapes are characteristic features on geologic maps of the Coastal Plain, and their sandy remnants are found far inland near towns such as Walterboro, Marion, and Kingstree. Using ocean soundings, barrier island chain remnants have also been mapped underwater many miles off the coast on the continental shelf.

STRUCTURE OF A BARRIER ISLAND

A barrier island system is made up of several interdependent parts: beaches, dunes, spits, barrier flats, channels, lagoons, salt marshes, inlets, and ebbtide and floodtide deltas. Young barrier islands may only have a single beach ridge and sparse vegetation, but mature islands can have several ridges and established, diverse plant communities.

A beach is an accumulation of sediments that runs from approximately 30 to 40 feet seaward of the breaker line to the farthest point onshore reached by the waves and their swash. The beach is the most dynamic part of a barrier island, for it is the source of sand for the entire island and is the site of the greatest movement and change. From the beach the island grows—fed by the waves and currents. It is here, on the beach, that a great deal of island life can be seen—including many types of seaweeds, worms, shellfish, crustaceans, arthropods, and birds.

The beach is divided into distinct parts: the inshore, foreshore, and backshore. The inshore is that part of the beach that is normally hidden by waves from the breaker line to about 40 feet offshore. In this zone the sands of the island are moved into bars, runnels, and ridges according to the tides, currents, seasons, and storms. From the inshore the sands of the island are replenished after a storm. The foreshore is that part of the beach visible between low and high tide where the surf and the swash break and run up the shore. The foreshore is the part of the island that most sunbathers, shell collectors, and walkers

Beach ridge on Folly Island

know best. The backshore is that part of the beach affected by seawater only during storms or exceptionally high tides. The backshore receives and holds the seed-containing seaweeds and grasses that are brought ashore during storms. Over time, the island flora can be extended here through natural propagation. The backshore ends at the dune line.

Sand is brought to and moved along the beach by the action of waves and swash. When waves hit the beach at an oblique angle (greater than 90 degrees), the sand particles are carried down the beach to accrete—perhaps in a beach terrace, called a berm. The waves may also continue down the island to help form a spit (the growing tip of the island, usually found at its end) depending on currents and the presence or absence of inlets. An island can often maintain the balance between sand erosion and sand accretion if left to adjust and shift its sand load in

Human attempts to hold the sand on Folly Island include groins and revetments. This groin is made of granite riprap.

natural ways over time. During the spring and summer months currents move sand from south to north along the coast. But the stronger fall and winter currents, combined with storms from the northeast, move more sand toward the south. Generally, sand removed from the island during storms will be returned to the

Beach erosion on Daufuskie Island (courtesy: Richard Lacy)

beach again in a matter of a few months by fair weather waves and the longshore current. Human-made "beach enhancers" such as jetties, groins, dams, and seawalls interfere with the normal sand replenishments of the beach. Rather than helping the island, these short-term devices can help cause greater erosion of a barrier island system.

The Dunes

The dunes on an island are formed mostly by sand that is blown up from the beach in the prevailing sea breezes during the day. To a lesser extent, the land breezes that blow seaward at night also contribute to dune building. Together, these breezes sculpt the dune ridge into its high profile. Plants, such as Atlantic beach grass and sea oats, eventually take root and hold the sand in place—stabilizing the dunes. When the dunes move, they move as a system with their protective plants intact, for these plants grow up out of the sand when buried and can migrate with the dunes. The plants cannot survive trampling and uprooting, however. If they die, the wind is free to blow away the sands. Therefore, the careless treatment of the dune vegetation—by walking on them and driving vehicles over the plants—can lead to the destruction of the dunes themselves.

The Barrier Flat

Directly behind the dune ridge is the barrier flat, which, because it lies in the lee (or shelter) of the dunes, is protected from most of the direct salt spray. As a result, plants, and even trees such as bayberry, oak, holly, and palmetto, can grow. Although pruned by the salt spray, the bigger trees protect the less salt-tolerant undergrowth. In turn, this undergrowth holds the thin soil under the trees in place and helps to provide a favorable microclimate for the bigger trees. The barrier flat, if vegetated, also provides a habitat for many diverse animal species such as birds, deer, reptiles, and small mammals—all of which enrich the ecosystem of the barrier island. If the undergrowth or the large trees are damaged, often neither can survive, and the island will be more open to wind and water erosion as a result.

The Salt Marsh

During storms, waves often cut through the dunes and carry large amounts of sand across the island. This sand, called an overwash fan, is deposited either on the barrier flat or on the shore of the lagoon between the island and the mainland. It is upon this new sand deposit that halophytes, salt-tolerant plants, can grow. At the lowest part of the marsh, where salinity reaches 0.7–3.2 percent, the spartina grass grows. At the lower high marsh, where salinity levels of 0.3–3.0 percent are found, such plants as sea oxeye, salt grass, and glasswort can survive. At the highest level of the upper high marsh, salt-meadow cordgrass, marsh elder, and sea myrtle grow. With new storms, particularly hurricanes, more sand is deposited, more marsh is established, and the island migrates landward. The back-barrier zone is therefore a geologically transient area that will eventually fill with clay, silt and organic material from the rivers, and coarser sands from the seaward side of the island. The marsh grasses trap fine sediments that come into the lagoon from rivers. The resultant environment is extremely nutrient-rich, which enables the marsh to become a nursery for hundreds of species of fish, shellfish, and crustaceans. This environment also provides feeding and nesting grounds for innumerable birds, reptiles, and other animals. If the marsh is not rejuvenated by overwash sands every 20 to 30 years, however, it becomes less healthy and less capable of maintaining life. When humans place buildings and seawalls to prevent overwash, they also prevent the necessary replenishment of the marshes, which leads to the slow decline of the life of the barrier islands.

BARRIER ISLAND GROWTH AND EROSION

A barrier island can only survive the battering of waves for as long as it has an adequate sediment supply. Sand becomes available by way of a littoral drift, which is the action of the longshore current that moves sediments along the coastline generally from north to south. Several factors, however, can influence whether or not a barrier island grows, remains stable, or erodes. The first factor is the availability of sand. With the damming of the Santee River in 1942, the sediment that was originally deposited at the Santee River Delta and moved down the coast to replenish the barrier islands, now flows to the bottom of Lake Marion, Lake Moultrie, and Charleston Harbor. As a consequence, during the last 40 years, the Santee Delta has lost 900 feet of shoreline. Partly as a result of this sediment starvation, 75 percent of South Carolina's barrier islands are eroding. Other factors may include rising sea levels and other human interferences with sediment movement along the coast. Directly south of the Santee River Delta, Bull Island, Capers Island, and Dewees Island all show signs of severe erosion.

Some barrier islands manage to maintain adequate amounts of sand due to natural reserves that accumulate seaward of the inlets along the coast. Created as storm waves cut channels through the narrow islands, these inlets, along with tides, become the chief means for building sediment reserves for the island chain. As high tides occur, seawater is pushed through the barrier inlet into the lagoon, carrying some sediment with it. Then as the water slows, much of the sediment is dropped and forms a floodtide delta. As the tide falls, much of the sediment that forms the floodtide delta is carried back out of the inlet because the force exerted seaward by the water as it moves out of the lagoon is greater than that coming in. The falling tide pulls large amounts of sand back out of the inlet into the ocean and forms offshore what is called the ebbtide delta. The ebbtide deltas can grow and shift overnight, and sometimes cause boats and ships to ground and sink. For the barrier islands, however, the ebbtide deltas provide sand that is reworked by the currents and accreted to the beaches, and sent down drift to replenish neighboring islands. Depending on their size and location, these shoals can refract waves, bending them in such a way that they send nourishing sand past an island. Then, as time passes and the ebbtide deltas shift position, the resumption of sand transport rebuilds the eroded island. It is estimated that some ebbtide deltas contain as much sand as the entire barrier island off of which they lie.

Charleston Harbor illustrates the effect of human interference with the movement of sediment along the coast. In 1898, a jetty system was constructed along the mouth of the harbor to stop the formation of unpredictable ebbtide shoals that disrupted the ever-increasing ship traffic in and out of Charleston. As a result, Sullivans Island and the Isle of Palms, located north of the harbor and the jetties, began to grow. Some of the sand that normally would have gone down the coastline was stopped and deflected to their beaches instead of continuing down the coast. On the other side of the jetties, however, Morris Island and Folly Beach, cut off from the sand by the jetties, began to erode. One result of the erosion was the loss of the lighthouse to Morris Island, which was onshore in 1935, but is now 1,600 feet offshore. Other islands farther south of Charleston are varied in their stability—for instance, Kiawah Island is accreting while its neighbor Seabrook Island is unstable.

On the beach at Edisto Island there are remnants of mud flats and oyster shell beds that were once on the lagoon side of the island. They have been exposed on the beachfront side as the island has migrated inland due to rising sea levels of the past 200 to 300 years. These deposits provide dramatic evidence of the landward migration of the barrier islands.

Farther down the coast at Hunting Island, erosion is occurring at a rapid rate. Over the past 125 years, the island has lost between 985 and 2,460 feet to erosion. Those who built the Hunting Island lighthouse in 1875 were perhaps wiser than many of today's developers. Knowing that barrier islands erode, they built the lighthouse with the capacity to be dismantled and moved, which they were forced to do by the encroaching sea in 1889. It is estimated that 300,000 cubic yards of sand, or 9 to 10 yards of beach front, are currently lost each year. In a determined effort to maintain the beach, the state of South Carolina has replaced lost sand on Hunting Island every few years. Other beaches along the coast—including Myrtle Beach, Folly Beach, Edisto Beach, and Seabrook Island—have been renourished

Hunting Island renourishment —pipeline depositing sand slurry onto Hunting Island beach, 1991 (courtesy: Tim Kana)

Hunting Island re-nourishment: pipe-line carrying sand from natural offshore deposit to renourish Hunting Island shore, 1991 (courtesy: Tim Kana)

Hunting Island beach renourishment program (courtesy: Tim Kana)

through a variety of federal, state, and private funding. The latest beach renourishment at Hunting Island occurred in March 1991 when 750,000 cubic yards of sand were dredged from a deposit offshore and piped onto the beach north and south of the lighthouse. Although the transplanted sand is ultimately carried away by currents, it will last for 5 to 10 years.

Bird Island shoal in Bull's Bay, a brown pelican rookery (courtesy: Tim Kana)

The Coastal Barriers Resources Act was passed in 1983 as an attempt to remove some of the incentives for developers to build indiscriminately along the coast. Previously, the federal government had offered low-cost flood insurance to those who would build on the islands—giving subsidies for the building of roads and sewers. All of this attracted development that the islands clearly could not sustain. The Act now makes it less financially attractive to develop these islands, and sets aside protected zones where building is not allowed. The South Carolina islands and inlets protected by the Act include Daufuskie Island, Dewees Island, St. Phillips Island, Harbor Island, Otter Island, Edisto Island, Morris Island, Waites Island, Bird Key, Pawley's Inlet, Captain Sam's Inlet, Litchfield Beach, and Debidue Beach.

South Carolina is very fortunate to have its coastal island chain nearly intact. Many other states along the eastern seaboard have lost scores of their islands and beaches to short-sighted management, overdevelopment, and a failure to understand the natural processes of this landform. Although some of South Carolina's islands are also threatened by over-development, erosion, pollution, and unwise management, many people and agencies are working to protect them. There yet remain thousands of acres of unspoiled coastal environments to study and enjoy.

Interesting Places to Visit at the Coast

Cape Romain National Wildlife Refuge—U.S. Highway 17 south of Georgetown, Moores Landing Pier on Bull Island Road in Georgetown County; call 803-928-3411—visits to Bull Island by boat.

Pinckney Island National Wildlife Refuge—Hilton Head Island, off U.S. Highway 278 past Graves Bridge, in Beaufort County—4,052 acres of island marsh environment.

Savannah National Wildlife Refuge—U.S. Highway 17 south of Hardeeville in Jasper County.

State Parks

Edisto Beach State Park—S.C. Highway 174, off U.S. Highway 17, in Charleston County—pristine salt marsh, fossil and shell collecting, nature center.

Hunting Island State Park—U.S. Highway 21, 16 miles east of Beaufort, in Beaufort County—nineteenth-century lighthouse, recently renourished beach, maritime forest.

Huntington Beach State Park—U.S. Highway 17, 3 miles south of Murrells Inlet in Georgetown County—salt marsh, maritime forest.

Myrtle Beach State Park—U.S. Highway 17 in Myrtle Beach, in Horry County—broad beaches.

Heritage Trust Preserves

Capers Island Heritage Preserve—off U.S. Highway 17 south of Georgetown, in Charleston County—2,100 acres; barrier island, reachable only by boat.

Tom Yawkey Wildlife Center—off U.S. Highway 17 in Georgetown County—17,700 acres of barrier island, maritime forest, pine flatwoods; by appointment.

Bellefield Nature Center—U.S. Highway 17 north of Georgetown, in Georgetown County—guided tours of estuarine and uplands environments.

Turtle Island Wildlife Management Area—between Daufuskie Island and Savannah (boat access only), in Jasper County—1,700 acres of salt marsh, maritime forest, and barrier beach.

Sullivans Island—off U.S. Highway 17 outside Mount Pleasant, in Charleston County—good beaches on an accreting barrier island.

Intracoastal Waterway—winds the length of the coast through the barrier and sea islands—fossil sites where Pleistocene sediments are seen at the surface or near-surface, provides the best view of many islands and marshes.

Myrtle Beach—U.S. Highway 17 in Horry County—over 30 miles of broad beaches.

University of South Carolina—Beaufort Campus, Coastal Zone Education Center, S.C. Highway 46 off U.S. Highway 278, in Beaufort County—field studies to barrier islands, speakers for Beaufort County schools.

Woods Bay, Clarendon County

CHAPTER 7

CAROLINA BAYS

Although the eastern coast of America is pocked with thousands of mysterious elliptical depressions, most are found within North and South Carolina—thus, the name Carolina bays. These Carolina bays lie along 800 miles of Coastal Plain from Maryland to northern Florida and are scattered over unconsolidated sediments from the Fall Zone to the coast. These bays are geomorphic features of great interest to geologists, biologists, and ecologists for they provide important clues about the environments that existed during the Quaternary Period along the coast of the eastern United States. In addition, they provide unique natural laboratories for the study of rare and endangered plant and animal species.

The Carolina bays are shallow, largely elliptical depressions that have long axes aligned in the same general northwest/southeast direction. Some bays are partly bounded by sandy ridges that are usually more pronounced on the southeastern side. Although some bays are completely surrounded by a rim, most have no rim at all. The majority of bays are found in the middle Coastal Plain of South Carolina and southeastern North Carolina, but there are some in South Carolina's upper and lower Coastal Plain as well. The bays range in size from 3 acres to thousands of acres, and in length from a few hundred yards to several miles. A small number of bays form permanent lakes, some of which overlie deep accumulations of organic soils, including peat. Some bays form bog swamps, while others stand as dry savannahs for most or all of the year. Their shapes are intriguing—many bays overlap others, some are found completely within others, and some are found in clusters.

Cypress trees in early spring at Woods Bay

Bald cypress in Cathedral Bay, Bamberg County, which is dry in late summer

Geologists of the 1930s over-calculated the number of Carolina bays that existed in North and South Carolina. It is now
thought that the early figure of four hundred thousand was highly
exaggerated. A recent survey conducted by the South Carolina
Non-Game and Heritage Trust estimates that there are only about
four thousand bays in the state. The bays are not evenly distrib-
uted across the Coastal Plain, but rather they range in frequency
from 3 in Chesterfield County to 410 in Horry County to none
at all in Beaufort County. The three county-cluster areas in which
bays occur with the highest frequency are:

1. Aiken, Barnwell, Orangeburg, Allendale, and
    Bamberg Counties
2. Sumter and Clarendon Counties
3. Darlington, Dillon, Marion, Horry, and Marlboro
    Counties

Number of Carolina Bays Per County

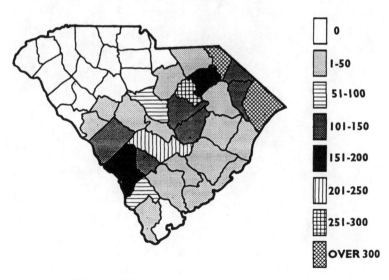

Number of bays per county in South Carolina (courtesy: South Carolina Department of Natural Resources, Non-Game and Heritage Trust)

Today, most of the bays have been drained by humans, their timber cut, and their rich boggy soils cultivated, so that their original outlines are often only dimly seen from the air. Evidence of a drained and cultivated bay can sometimes be detected from the ground because its soil is generally the darkest and most productive in an area. These areas are often in sharp contrast to the surrounding Coastal Plain soils, which are poor in quality and composed largely of sands. Out of the thousands of original bays, only 219 have been spared the destructive alteration of agriculture, urbanization, highway construction, and logging. Of these, only thirty-six remain close to their original condition. Because many bays fill and dry out seasonally, they were not considered in the past to be permanent wetlands—even

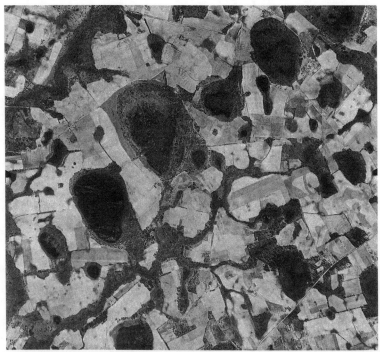

Allendale County bays, 1943, showing many near-pristine bays (courtesy: University of South Carolina, Thomas Cooper Map Library)

though it was known that their existence helped the survival of many plant and animal species dependent on water at least part of the year. As late as the 1970s, the federal government subsidized farmers to drain the bays and plant soybeans. In the rush to utilize the bays, most of their rare and beautiful habitats were permanently lost. Today, however, the remaining bays are at least partially protected by the 1972 Federal Clean Water Act.

Some bays are boglike ecosystems that constitute natural swamp oases for varied and rare flora and fauna. Many concerned citizens feel that they should be better preserved. In 1983, the South Carolina Non-Game and Heritage Trust made the protection of the bays a priority and launched a five-year project to identify and study them. Since that time, several bays have been purchased by the Heritage Trust so that they will be left for the future in as natural a condition as possible.

BAY FORMATION THEORIES

The Carolina bays were first described in the early eighteenth century when English naturalist John Lawson referred to them as "pocosins"—the Algonquin Indian word for "swamp on a hill." These pocosins were later named "bays" after the profusion of bay tree varieties that grew in them. Inquiry into the geological origins of the bays began when South Carolina state geologist Michael Toumey described the bays in 1847 as looking like "racetracks" after noting their high, circular sandy rims. He wrote that the bays did not appear to be lime sinks, which are deep conical depressions, for those he saw were instead shallow depressions with gently sloping sides (Savage 1982, 14). Writers of the early twentieth century also mentioned the bays in their scientific literature. With the arrival of aerial photography in the 1930s, the full, dramatic extent of the bays was seen for the first time, and several formation theories were offered to the public. These included the fanciful theories of spawning fish and frolicking sea monsters, both of which, it was proposed, scooped out depressions at the bottom of the ocean that were later exposed as sea levels dropped! Today geologists and geographers, while discounting the most fanciful theories, still debate the origin of the bays. Three theories of bay formation that have been seriously considered include meteorite bombardments, tidal eddies, and wind deflation in combination with lake wave scouring. The various hypotheses provide a fascinating view of the processes by which scientists investigate geological enigmas such as those provided by the Carolina bays.

The meteorite theory is the most exciting, but the least scientifically convincing. Perhaps influenced by the outer space rage that swept the nation throughout the 1930s, two geologists from Oklahoma named Melton and Shriever proposed in 1933 that the Carolina bays were created by meteorites. They proposed their theory after viewing the first aerial photographs of the bays. They wrote:

> Aerial photographs of a district on the Coastal Plain of South Carolina reveal hitherto unknown relationships among surface depressions of a peculiar type the origin of which has long been a subject of speculation. These relationships include (1) a smoothly elliptical shape, (2) parallel alignment in a southeastern direction, (3) a peculiar rim of soil, which, with unimportant exceptions, is invariably larger at the southeastern end than elsewhere, and (4) mutual interference of outline. Consideration of all these facts leads to the conclusion that the origin is not directly attributable to ordinary geologic processes. On the contrary, a hypothesis involving impact by a cluster of meteorites is presented as the most reasonable explanation. (Kaczorowski 1977, II-20)

Geologist W. F. Prouty promoted and expanded the meteorite theory, and the sensationalist journalism of the day eagerly publicized the possibility of an ancient extraterrestrial impact in South Carolina. Journalists published articles promoting the catastrophic origin of the bays to the public so convincingly that the idea is still widely believed and repeated. On close examination, however, the meteorite theory can easily be discounted because it offers no evidence that points to an extraterrestrial source for the formation of the bays.

While at first glance the shape of the bays might resemble the craters of ancient meteors, there are no meteorite fragments associated with Carolina bays either at the surface or at depth. There are also no shocked quartz sand grains, which would be expected from the heat and concussion of such a powerful impact. Seismic reflection studies have shown no deep disturbance of the bedrock underlying the bays. There are no ejecta breccia (fragments of fused rock) on the rim of the bays—there is only sand. Meteorites can travel thousands of feet under the surface as they explode on impact—but the bays are shallow. Perhaps most important, due to the high speed of meteors (which travel at 50,000 miles per hour), their impact causes great explosions as they hit, which creates round, or nearly round, craters—not oval or elliptical ones. In fact, no known meteor craters are elliptical. Craters found in the United States (Iowa, Colorado, and Arizona), Mexico, Canada, and on the moon are all round or nearly round. Given the lack of scientific evidence that might point to a meteorite shower as the cause of the Carolina bays, most geologists rate the theory low on a scale of probability. It is important to note that the bays are found only on the unconsolidated sediments of the state's Coastal Plain. Because the Carolina bays are not found in the hard rock of the Piedmont, the evidence points toward a tidal or eolian genesis.

In 1954, geologist Wythe Cooke (*Carolina Bays and the Shapes of Eddies*) offered a theory that natural tidal eddies created the bays as ocean currents moved underwater near the sites of the various Pleistocene coastlines. He proposed that as the sea receded, the prevailing winds and waves assisted in building up the height of the bay sand rims. Because eddies naturally form between two opposing currents, between a current and an obstruction, or as a current enters quiet water, Cooke felt that the bays could have evolved from noncatastrophic processes in inland estuaries or on tidal flats when the Atlantic Coastal Plain was still covered by the shallow sea.

Cooke was very much taken by the remarkable elliptical shape of the bays and their parallel orientation that consistently points in a general northwest-southeast direction. Cooke based his theory principally upon these characteristics. He proposed that tidal eddies or swirling currents responded gyroscopically to the rotation of the earth and formed elliptical shapes with their long axes oriented close to 45 degrees west in the Northern Hemisphere. This positioning would be at a proper angle for South Carolina's latitude. (The long axes of the bays would point northeast/southwest in the Southern Hemisphere.) The larger the bay, the more elliptical it would be, for it would have more surface area to respond to the rotation of the earth. Most small bays are indeed round. Most large bays are nearly elliptical, and some do achieve elliptical perfection.

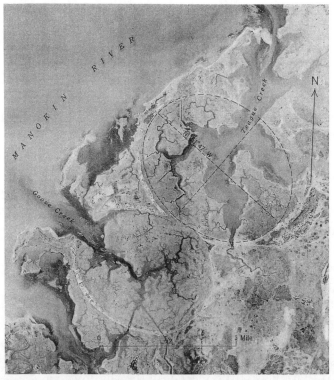

Plate 43, Cooke. "Elliptical Ridges at Rumbley, Somerset County, Maryland" (courtesy: U.S. Geological Survey)

Cooke wrote: "Many bays exhibit such perfect parallelism and remarkable symmetry that randomness in their formation must be excluded. . . . The Carolina Bays have the proportions and orientation approximating those of the ideal fixed eddy" (*Carolina Bays and the Shapes of Eddies,* 1954). According to Cooke's analysis, such tidal eddies are today forming future Carolina bays along the Atlantic coast in tidal estuaries in such areas as Chesapeake Bay.

Cooke's theory has few adherents among today's geologists, because few believe that baylike features can be seen forming in inlets or along the coast, or that the mechanisms described by Cooke are valid. If the theory of the oceanic origin of the bays is true, then the ages of the bays should correspond to the ages of the terraces on which they lie. The dating of the bays remains unfinished and, as a consequence, no age correlation has been found to match the age of the bays with those of the terraces. The few that have been dated are all Pleistocene or younger in age ranging from 6,000 to 40,000 years with estimates of up to 250,000 years. There are bays, however, which lie on Pliocene terraces, but they remain undated.

The formation theory that presently enjoys the strongest academic support is that proposed by Raymond Kaczorowski, who believes that the bays are natural depressions deepened and expanded by wind and water. He points to Carolina bay analogues in the permafrost thaw lakes of the Alaskan Arctic and to playa lakes in Tierra del Fuego, Chile. Both of these areas have strong prevailing winds and both have lakes that are oriented perpendicular to the direction of those winds. The mechanisms for the formation of these distinctive lakes include the internal, wind-driven currents of the lakes (or eddies) that scour their sides. Kaczorowski proposed that the Carolina bays were formed in much the same manner ("The Carolina Bays and Their Relationship to Modern Oriented Lakes," 1977).

Recent research into paleoclimates suggests that the conditions for strong prevailing winds might have existed at South Carolina's latitude during the Pleistocene, which was the time

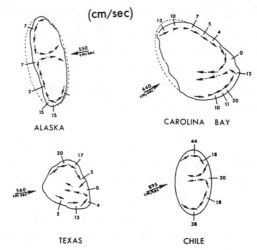

CURRENT VELOCITIES AND CIRCULATION
(cm/sec)

ALASKA

CAROLINA BAY

TEXAS

CHILE

Current velocities and circulation. (Credit: Raymond
T. Kaczorowski, "The Carolina Bays and their Relationship to Modern Oriented Lakes," 1977)

of formation of all the dated bays. It was a time when sea levels
around the world rose, as the recurring glaciers of the epoch
melted, and then fell again as the glaciers reformed. The fossils
of fauna and flora give evidence that, during glacial periods,
the climate of South Carolina became colder and drier. That
climate would have invited the formation of lakes. Vegetation
would have been absent or very thin along the coasts with dune
fields, scrub, and grasslands. High pressure over the glacial north
could have increased the force of the prevailing winds over
South Carolina from the west and the southwest, resulting in
oriented lakes (bays) aligned in a northwest-southeast direction.

Some geologists have proposed that the bay sand rims were
created as ice push ridges. They believe that as the lake thawed,
large pieces of ice acted like battering rams as they were thrown
against the shores on the eastern half of the bays by the cold
Pleistocene winds. As the ice hit, it shoved the sediments into a
high ridge that can be seen today as the sandy rims found on
some bays.

Sand rim in Woods Bay

STRUCTURE AND ECOLOGY OF THE BAYS

The Carolina bays in North and South Carolina contrast sharply with each other. The majority of South Carolina bays are dry for most of the year, with their water levels rising and falling with the seasons. Only a few bays in Marion and Horry Counties contain peat. Most of the South Carolina bays rest on top of an impermeable clay layer that can be as thick as 25 feet. In many bays, this clay traps the water level above the water table, creating permanent or semipermanent ecosystems that support the myriad bay fauna and flora. Many swamp bays are encircled by a sand rim 5 to 15 feet above the surface of the swamp, which can extend out 200 to 300 feet. The bay depression is shallow, deepest at its southwestern quadrant, and approximately 20 to 25 feet below the surrounding ground level. The bog itself is shallower still, not exceeding 15 feet deep and usually only 3 to 4 feet deep. The shallowness of the bays creates an oxygen-rich water environment that supports extensive vegetation such as cypress and gum trees. The clay-based bay can support a much greater diversity of life than the peat-based bays, for the clay contains and traps more life-supporting nutrients and is less acidic.

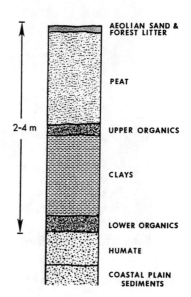

Diagram: Generalized bay stratigraphy (Credit: Raymond T. Kaczorowski, "The Carolina Bays and Their Relationship to Modern Oriented Lakes," 1977)

The bays in North Carolina are primarily of a pond pine woodland type containing deep peat deposits. Commonly known as pocosin bays, they are often underlain by a highly impermeable eight-to-twelve-inch hardpan made up of sand and clay, and cemented by a relatively insoluble material such as iron to help assure the containment of water. The peat bays were formed from mildly decomposed organic material, including large amounts of sphagnum moss. As the material sank to the floor

Virginia bay (sweet bay) in Woods Bay State Park

Pond cypress in bay near Sumter County (courtesy: South Carolina Department of Natural Resources)

of the bay lake, it was deposited in a highly anaerobic (or oxygen-poor) environment. This retarded its decomposition and allowed the peat to build to great thicknesses. Today, the acidic peat is in high demand for use by gardeners, and is harvested from some bays in North Carolina and from some nonbay bog wetlands in South Carolina. The low-oxygen, acidic environment of the peat bays is habitable by a small variety of plant life—including various species of blueberries, wild azaleas, stunted loblolly bays, hollies, and pond pines. The area is abundant, however, in vines and briers. The pocosin peat-rich bogs are rimmed with sedges and grasses, and mats of moss float on the surface of the water. The famous flora of the Carolina peat bays—Venus fly trap, pitcher plant, and sundew—evolved into rare insectivores. Because the peat bogs are so low in nutrients, these plants developed the ability to catch live prey in an attempt to augment their food supply.

A number of the endangered and rare plant species which exist in South Carolina are found only in the bays. Even dry bays contain rare species, such as mock bishop's weed, found in the depression meadow bays of the Peach Ridge in Edgefield County. Other species nurtured by the bays include quill-leaf, rose coreopsis, and spoonflower—three out of more than twenty-three endangered species in the bays. Rare animals that are protected by the bays include bobcat, osprey, many amphibians, and, in Horry County, the black bear.

Lewis Ocean Bay in Horry County, a pocosin bay (courtesy: South Carolina Department of Natural Resources)

Diagram: Section and plan views of a typical Carolina bay, indicating key morphological features, soil profiles, and vegetation types, modified from Buell, 1946 (by R. R. Sharitz; courtesy: U.S. Fish and Wildlife Service.)

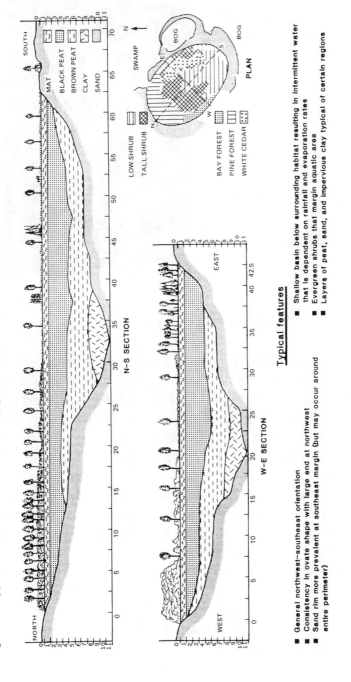

Typical features

■ General northwest–southeast orientation
■ Consistency in ovate shape with large end at northwest
■ Sand rim more prevalent at southeast margin (but may occur around entire perimeter)
■ Shallow basin below surrounding habitat resulting in intermittent water that is dependent on rainfall and evaporation rates
■ Evergreen shrubs that margin aquatic area
■ Layers of peat, sand, and impervious clay typical of certain regions

The Venus fly trap found in pocosin bays of South Carolina
and North Carolina (courtesy:
South Carolina Department of
Natural Resources)

WOODS BAY STATE PARK

A fine example of a Carolina bay has been preserved as the
Woods Bay State Park near Olanta. The park is one of the most
accessible in the state—off U.S. Interstate 95, southwest of Florence. It contains a stunning 1,548-acre bay that was saved from
alteration by logging in 1972 and was turned into the state's
first totally natural park. In the 1920s, most of the cypress trees
in the bay were cut using tramways to haul them from the interior of the swamp. In 1971, logging companies were preparing
to cut again, but naturalist J. C. Truluck led a movement to
preserve the bay as a state park. With the aid of the state legis-
lature, the South Carolina Department of Parks, Recreation, and
Tourism, and other supporters, Truluck was successful in pre-
serving Woods Bay. In stark contrast to the pristine beauty of
Woods Bay, across the road lies its less fortunate twin, Dial
Bay, which was put to agricultural use in 1971 and now lies
drained and dry.

Dial Bay, turned to farmland

The varied habitats of Woods Bay contain both swamp and savannah, as well as sand rim and pine barrens. In the interior of the bay, a cypress-tupelo community can be found with pond cypress, swamp tupelo, red maple, sweet gum, Virginia willow, wax myrtle, and ferns. It is a refuge for dozens of varieties of birds—including the yellow-billed cuckoo, osprey, hawks, owls, little blue herons, and songbirds of every description. Alligators and other reptiles join bobcats, otters, minks, muskrats, raccoons, grey foxes, and even black bears in making the near-pristine refuge of the bay their home. In the sand ridge community along the sand rim, wild turkeys can be seen wandering amid longleaf pines, turkey oaks, and prickly pear cacti. Wild turkeys have been reintroduced at Woods Bay and are making a successful comeback. The evergreen shrub community contains the flora that comprises the namesakes of the Carolina bays—the sweet bay, red bay, and loblolly bay trees. On the southern end of Woods Bay, the savannah or marsh community flourishes, and is the home of many bird species as well as various reptiles, amphibians, and mammals that live in its sedges, rushes, and panic grasses.

In addition to Woods Bay, several other South Carolina bays have been preserved and are open to the public. The South Carolina Department of Natural Resources publishes a *Heritage Trust Program Preserve Guide* on the bays describing in detail the features of each bay and how to locate them.

The South Carolina bay preserves include: Savage Bay near Camden in Kershaw County; Lewis Ocean Bay near Myrtle Beach in Horry County; Cartwheel Bay near Conway in Horry County; Bennetts Bay near Manning in Clarendon County; and Cathedral Bay near Olar in Bamberg County. In addition, there is a Carolina bay just outside of Little Pee Dee State Park near Dillon. In North Carolina, Singletary Lake State Park and Lake Waccamaw State Park both contain bays and are located not far north of the state line. These are good examples of pocosin bays.

Aerial view of overlapping bays (courtesy: South Carolina Department of Natural Resources)

The Carolina bays are special gifts from ancient interactions of land, water, and wind. Once gone, the ecosystems that they support cannot be reproduced. The bays are not as dramatic a natural resource as the South American rain forest, but, like the rain forest, the bays nurture and preserve rare forms of life when left undisturbed.

Archeocete tooth (courtesy: South Carolina State Museum)

CHAPTER 8

SOUTH CAROLINA FOSSILS

A large portion of the rocks and sediments of South Carolina contains fossils. Because many of the state's ancient environments were shaped by the interaction of the ocean and rivers with the land, the fossils in South Carolina are largely marine and estuarine. Fossils of land animals, including mammals, are preserved from the Cretaceous Period to the Recent Epoch in South Carolina and are found from the Fall Zone to the coast. At the southern end of the Carolina slate belt, trilobite fossils of much more ancient Cambrian Age are found and some petrified wood of undetermined age has been discovered in the Carolina slate and Charlotte belts.

New fossils are still being discovered and extensively studied by the South Carolina State Museum, the Charleston Museum, and the Smithsonian Institution. From the muds near Charleston, previously undiscovered species of whales, turtles, and dugongs have been unearthed, and the museums believe that many new species will continue to come to light in the Cenozoic rocks of the Coastal Plain. There is enough fossil digging to be done in South Carolina to keep a student of paleontology, professional or amateur, busy for many years searching for the secrets of life's development within the rocks and sediments. There are lots of fun ways for students and hobbyists to get involved. Opportunities for the exploration, digging, and discovery of fossils have never been more abundant. Curator Jim Knight of the South Carolina State Museum has described South Carolina as "a black hole of paleontology," meaning that fossil exploration in the state is only in its infancy and the possibilities for exciting discoveries are tremendous.

Curtis Bentley of the South Carolina State Museum prepares a recent crocodile fossil skeleton found near Summerville, 1990 (gavialosuchid, sp)

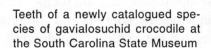

Teeth of a newly catalogued species of gavialosuchid crocodile at the South Carolina State Museum

Great white shark model (courtesy: South Carolina State Museum)

THE NATURE OF FOSSILS

Fossils are the preserved remains of animals and plants, or the traces left by them. There are many fossil forms including casts, molds, footprints, teeth, eggs, burrows, and nests. By collecting and studying these remains, both experts and hobbyists have contributed a great deal to the growing understanding of how and when plants and animals developed, how they lived and died, and how they finally became extinct.

The form that a fossil ultimately takes is determined by the environment into which an organism falls or is carried after it dies. If it falls into water and is quickly buried by sediments, it is preserved in an anaerobic environment. Without oxygen to destroy its tissues, time works with sediment and water to preserve the form of the animal or plant through such processes as mineralization or carbonization.

Mineralization occurs as tissues are replaced over time by minerals, such as silica or iron, which are in the water surrounding the organism. A well-known example of a mineralized fossil is petrified wood, which forms as silica replaces plant tissue over time. There are many deposits of petrified wood in the gravel pits and sand quarries of the Pee Dee area of Darlington County and along the Wateree River valley in Sumter County. Although the specimens are not as colorful as those found in the Petrified Forest of Arizona, they are still worth collecting. The Pleistocene fossils of Edisto Island are mineralized by phosphates, which hardens them and gives them a distinctly black color.

Carbonization creates dark-colored fossils as animal or plant tissues are reduced over time to one of their essential elements—carbon. Carbonized plant and animal remains can leave perfect lifelike imprints in stone. These remains are often found in shale, a fine-grained rock that forms by compaction and very little heat. When shale is split evenly along its layers, the fossils are revealed. There are a few clay beds in Aiken, Darlington, and Chesterfield Counties where plant fossils can be found in the form of leaf imprints.

Petrified log from Marlboro County at the South Carolina State Museum (courtesy: South Carolina State Museum)

Sometimes a fossil shell or bone dissolves completely while buried. Often the space that remains, called a mold, has the same shape as the original organism. A cast can form if the mold is later filled with minerals, sand, or mud. The same principle is used to build sand castles at the beach by adding wet sand to a bucket (the mold), and then tipping out the shaped sand (now a cast) to use as a building block. Shell fossils created by molds and casts are very common in South Carolina. The structures of these fossils lend themselves easily to preservation.

Traces are also very common fossils and include footprints, nests, and burrows. Often paleontologists see only the traces of animals and must infer from such evidence the nature of the animal that left the trace. In South Carolina trace fossils are very commonly seen in sandstone, such as the worm burrows found below the waterfall at Peachtree Rock in Lexington County.

Although fossils may appear at first glance to be drab and dull, many sparkle with dramatic minerals such as pyrite and calcite. The color of a fossil depends on where it formed and the minerals that were present in the sediments during its long years of burial.

Worm burrows at Peachtree Rock

If given an adequate amount of time, the minerals that leach through the rock and soil into the fossil can fill it with vibrant and exotic colors. For instance, iron can color sediments and fossils yellow and red, and copper can color them blue or green. There are several sites in South Carolina (such as the Giant Portland Cement limestone quarry in Dorchester County, and the Savannah River chert beds south of Augusta) where the collector can view unusual fossil and sediment color.

THE IMPORTANCE OF PALEONTOLOGY

Paleontology is one of the sciences that can be enjoyed purely for its own sake because fossils are both attractive and interesting. But the knowledge that paleontologists gain through their explorations of fossils is also profoundly useful. Their knowledge affects the information base of other sciences such as geology and evolutionary biology. For example, paleontological studies help scientists locate mineral and hydrocarbon deposits, including oil and coal. Through a microscopic study of sedimentary rock, foraminifera and other tiny evidences of life can be used to track hydrocarbon deposits. Index fossils, which are characteristic fossils that identify and date the rock strata in which they are found, help geologists follow coal seams and ore deposits within a local area, and even across oceans from one geologically related continent to the other.

Paleontology has provided vital knowledge about the development of life on earth over the past 3.5 billion years. Using marine fossils, scientists can determine the nature of ancient climates, or paleoclimates, by assessing the oxygen isotopes found in fossil shells. These isotopes can reveal the temperature of seawater at the time the shells formed. Such studies help to determine the nature of the world's climate during the various geological periods, the extent of ice caps, and even the movement of continents.

THE HISTORY OF LIFE IN
SOUTH CAROLINA

With the exception of a few Cambrian fossil sites in the Carolina slate belt and some petrified wood in the Piedmont, the visible fossils of South Carolina begin in the Cretaceous Period. This does not mean, however, that plants and animals from earlier periods never lived in the state. It means that their remains never fossilized or that traces of them have either been eroded away or buried deep under thousands of feet of sediment in the Coastal Plain and along the continental shelf. South of the Fall Zone, the Cretaceous Period and all of the epochs of the Tertiary and Quaternary Periods are represented and can be investigated by paleontologists.

The Precambrian Period

Life began in the Archean Eon (3.8 billion to 2.5 billion years ago). The evidence for this is contained in rocks from Greenland, South Africa, and Australia. Microscopic investigations of these rocks have shown that earliest life consisted of bacteria and blue-green algae. Some blue-green algae shaped themselves into small clusters and strands, but also mixed with trapped sediments to form large toadstool-shaped, tidal structures called stromatolites. These early plant colonizers of earth built up the oxygen in the atmosphere over time through the processes of photosynthesis. After they had taken in carbon dioxide and given off free oxygen for billions of years, the atmosphere and oceans contained a high enough concentration of oxygen to support larger life forms. As a result, during the Late Proterozoic Eon, shortly before 600 million years ago, there was an explosion of complex life. It developed in the sea as myriad species of soft-bodied marine animals. By the Early Cambrian, animals with fossilizable body parts had evolved (as seen in the famous Burgess Shales of Alberta, Canada) and most of the invertebrate phyla were in existence.

With the evolving atmosphere, marine animals became capable of forming hard body parts out of oxygen and calcium from seawater. With their new skeletal parts, these animals could better survive the rigors of life on the ocean bottom and so the stage was set for the development of a variety of life. As a result, throughout the millennia from the Late Precambrian to the present, millions of organisms have appeared, diversified, and become extinct.

The Paleozoic Era

The Cambrian Period

The Cambrian Period (570 to 505 million years ago) saw the development of all the major phyla of life. At this early stage, each phyla had very few genera. The most abundant hard-bodied animals of the period were trilobites that make up 60 percent of all fossils recovered from the Cambrian Period. Also abundant during this period were mollusks such as the brachiopods, which account for 10 to 20 percent of Precambrian fossils. Brachiopods were bivalves with a distinctive winged shape that lived throughout the Paleozoic Era—a few species are even alive today. Also common and diverse were the echinoderms, sponges, and calcareous algae.

The Ordovician Period

The Ordovician Period (505 to 438 million years ago) gave rise to the earliest fish with the emergence of Agnatha (the jawless fish related to the lamprey of today). Their phylum, Chordata, was the last of the phyla to develop. From the Chordata later sprang the cartilaginous and bony fishes and the amphibians, and much later reptiles, birds, mammals, and humans. The Ordovician also saw the widespread development of coral reefs, nautiloids, graptolites, gastropods, and bryozoans. Crinoids, which were a type of echinoderm resembling a stalked sea plant, colonized the shallow sea floor.

The Silurian Period

The Silurian Period (438 to 408 million years ago) saw the emergence of the first simple vascular land plants and the first winged insects. Scorpions, among the first terrestrial arthropods, began to settle on dry land. The fauna in the seas continued to diversify, especially the trilobites, graptolites, and brachiopods.

The Devonian Period

The Devonian Period (408 to 360 million years ago) was a time of transitions. The cartilaginous fish (such as sharks) and myriad species of *Osteichthyes* (the bony fish) spread and diversified. These fish were so successful that the Devonian Period is called the "Age of Fishes." The early amphibians evolved from the lobe-finned fish, called *crossopterygians,* and began to crawl onto land as the first terrestrial vertebrates. Huge fern-like plants also began to spread over the land. In the seas the ammonites, as well as the brachiopods and trilobites, flourished.

The Mississippian Period

The Mississippian Period (360 to 320 million years ago) witnessed the emergence of the first reptiles, descendants of the amphibians. These tough, adaptive creatures were able to take advantage of dry areas that the amphibians could not. The reptiles developed a tough and leathery outer skin layer and an eggshell that allowed them to move away from water to survive. They were therefore able to carve out a niche for themselves in the drier regions of the world. Very prevalent in the seas during this period were crinoids and blastoids.

The Pennsylvanian Period

The Pennsylvanian Period (320 to 286 million years ago) was the time of immense swamps with vascular plants such as giant ferns, horsetails, and conifers. These swamps were to become the great coal fields of the world after the plants died, fossilized, and were compressed for millions of years. The amphibians were still among the prominent animals, for the climate was wet and warm. They had plenty of food, and the reptiles were too underdeveloped to provide much competition—a situation that changed during the Permian Period.

The Permian Period

The Permian Period (286 to 245 million years ago) witnessed a great extinction. The world's climate became drier toward the end of this period as the individual continents continued to move together to form the supercontinent of Pangaea. This merging of continents caused the uplifting of the land, and the reduction in size of many continental shelves. Sea levels changed, ocean currents shifted, and glaciation occurred in the Southern Hemisphere. Paleontologists now believe that as the seabed rose in some areas and fell in others, many of the invertebrates lost their habitat. In addition, the overcrowding of the continental shelves and the subsequent increase in competition for food and living space contributed to the large-scale, rapid extinction of fauna. Those species that were lost included the trilobites, which had been in existence since the Late Precambrian; most of the ammonites; and many of the corals, brachiopods, and tiny foraminifera. Fifty percent of the species then in existence disappeared. The therapsids, however, the mammal-like reptiles, survived. From the descendants of these reptiles would come the mammal explosion of the Cenozoic Era.

The Mesozoic Era

The Triassic Period

The Triassic Period (245 to 208 million years ago) was the earliest period of the Mesozoic (middle life) Era. The Late Triassic Period saw the break up of Pangaea, the supercontinent that had lasted for nearly 100 million years. Along with this break up, the relatively mild world climate changed. The crustal stretching, downfaulting, and rifting (along with volcanic activity and the ultimate rise of sea levels) had a profound effect

on animal and plant life. For example, the rifting of the continents built undersea mountain ranges of lava which displaced
such large quantities of water that sea levels began to rise all
over the world. Rising sea levels affected both climate and habitat. The first primitive mammals began to develop during this
period, but they were small and barely noticeable. The reptile
group remained the dominant species on land for the next 180
million years. From the reptiles emerged not only the mammals
and the birds, but also the great dinosaurs.

The Jurassic Period

During the Jurassic Period (208 to 144 million years ago),
Pangaea completed its break up, and the world's flora and fauna
became isolated because of the shifting continents. This isola-
tion caused the animals and plants of the time to change in very
different ways. The mammal group diversified and produced
monotremes, marsupials, and placentals. Birds evolved and the
amphibians declined.

The Cretaceous Period

The Cretaceous Period was the last of the Mesozoic Era (144
to 66.4 million years ago), and it is the first period, after the
Cambrian, whose fossils are found in South Carolina sediments.
Three very important types of flora developed during the Cretaceous: the deciduous plants, the flowering plants, and the
grasses. Their appearance at this time helped make possible the
later explosion of mammal life during the Cenozoic because
these flora served as a plentiful and highly diversified food
source for animals of every kind.

The Cretaceous Period was the finest hour for the dinosaurs,
but it was also their last—they became extinct at the end of it.
Along with the dinosaurs, 75 percent of all species alive at the
time became extinct. The reasons for the great extinction are
still unknown; however, various hypotheses range from global
warming to disease to comet or asteroid impact, and researchers are still energetically studying this question in search of
more definitive answers.

It is now evident that many species of dinosaurs lived in the South. In 1989, Dr. David Schwimmer, a geologist from the University of Georgia, announced the discovery of the first dinosaur fossils ever found in Georgia. He located bones of several hadrosaurs and tyrannosaurs in upper Cretaceous strata near Columbus. Earlier, in 1986, amateur paleontologist Aura Baker first discovered dinosaur remains in South Carolina. While digging 5 feet down in river sands near Kingstree, she unearthed two hadrosaur teeth that measured just over ¼ inch in length. Although some species of hadrosaurs grew to be 30 to 40 feet in length, they had small teeth because they were vegetarians. During the lifetime of a hadrosaur, its mouth was filled with thousands of tiny teeth that fell out and were replaced as it fed on very tough vegetation—including evergreen bark, cones, and needles. These teeth found at Kingstree, unfortunately, cannot be used to estimate the hadrosaur's length, and its exact identity remains a mystery.

The first carnivorous dinosaur fossils found in South Carolina were discovered in Quinby, near Florence, in 1992. A tooth about 1½ inches long, believed to be a dromaeosaurid, was found. Then, in 1994, three more dromaeosaurid fossils were located in the Black Creek Group in the Donoho Creek Formation at the

South Carolina's first dinosaur fossil teeth—from a hadrosaur found in 1987 near Kingstree (courtesy: South Carolina State Museum)

Great Pee Dee River, also near Florence. These included an ingual (a claw core), a metatarsal bone, and a serrated tooth. The claw core is similar to, but is not from, a velociraptor. This dinosaur could have been 6 to 10 feet long. The other two fossils appear to be from separate carnivorous species. The search continues in the Black Creek Group for further dinosaur remains.

The Cretaceous was a period of success for animals other than the dinosaurs. The seas contain fossils of what must have been true sea monsters—the plesiosaurs and the mesosaurs, ocean-swimming reptiles over 30 feet in length. The twelve-foot-long giant sea turtle, *Archelon,* swam in the seas here as well as crocodiles and soft-shell turtles. Invertebrate fossils included ammonites, myriad species of mollusks (including oysters), and belemnites. The belemnites, relatives of the squid, have been valuable to scientific studies. By measuring the number of oxygen isotopes within their calcite inner shells, the nature of the paleoclimate at the time they swam in the seas can be determined. These isotopes have shown that they lived in a tropical ocean.

The Cenozoic Era

The present era is called the Cenozoic, meaning "recent life" (66.4 million years ago to Recent). The vast majority of South Carolina fossils belong to this era. Although brief, the Cenozoic has witnessed many dramatic developments in animal and plant life. With persistent exploration the collector can find many sites, old and new, in which fossils of the animals described in the following sections can be found. Most of the fossils found in South Carolina are of marine animals. It is important to remember, however, that above what is now the Fall Zone, in the Piedmont and Blue Ridge regions, many nonmarine animals flourished. Although conditions were not conducive to fossilization there, some day isolated remains may yet be found.

The Cenozoic Era is divided into two periods, the Tertiary (66.4 to 1.8 million years ago) and the Quaternary (1.8 million to present). These two periods are subdivided into epochs, which are time boundaries that delineate dramatic changes in fauna.

The Tertiary Period

The Tertiary Period contains the Paleocene (66.4 to 57.8 million years ago), the Eocene (57.8 to 36.8 million years ago), the Oligocene (36.8 to 22.7 million years ago), the Miocene (22.7 to 5.3 million years ago), and the Pliocene (5.3 to 1.6 million years ago) Epochs. It is by far the longer of the two periods.

The Paleocene Epoch

By the end of the Paleocene Epoch (66.4 to 57.8 million years ago), most of the modern orders of mammals had come into existence, including rodents and primates. With the extinction of the dinosaurs, previously held niches were open for exploitation by new animal species. The Paleocene mammals were small, so searching for them today involves a great deal of careful hunting and sifting. In South Carolina, the digging of the Santee-Cooper Rediversion Canal on the Santee River near St. Stephens in 1984 brought to the surface the earliest mammal fossil ever found in the state. A paleanodont was found, which was the precursor of the glyptodonts, armadillos, sloths, and anteaters. The site has long since been filled, but it gave tantalizing proof of the existence of early mammals in South Carolina.

The Eocene Epoch

The Eocene Epoch (57.8 to 36.8 million years ago) was a time of continental connections. Land bridges formed between North America, Greenland, and Europe, and between North America and Siberia. India welded itself to Asia and built the Himalayas in the process, and Africa connected with Europe and Asia. Land animals began to mix easily by crossing over these land bridges and the competition that resulted from this mixing caused some species to prosper and others to die out.

Primitive horses, tiny camels, and giant birds developed. In South Carolina, the Eocene fossil record represents largely that of life on shallow continental shelves, beaches, and estuaries— for most of the state was inundated by the sea during this time.

Fossil field found at Giant Portland Cement Quarry

Eocene ocean fauna included baleen and toothed whales (*Cetacean*). The archeocete *Basilosaurus cetoides* is a striking member of South Carolina's fossil whale family. It was a fish-eating, jagged-toothed, sixty-five-foot long, thirty-to-forty-ton proto-whale with a keen sense of smell. It is thought that it may even have given birth to its young on land like modern seals and walruses. Only South Carolina and New Zealand have whale fossils of all three existing orders: mysticete, odontocete, and archeocete. Porpoises, crocodiles, and a wide array of bottom-dwelling sea life, including nautiloids and sawfish, were also prevalent during this epoch.

The Oligocene Epoch

The explosive diversification of life that had occurred in the Paleocene Epoch began to slow during the Oligocene Epoch

(36.8 to 22.7 million years ago). The sea retreated, and another extinction of animals, largely marine, began. Marine fauna decreased from ninety-five families to seventy-seven families. The sea retreated during the Oligocene from the inundation of the Eocene as the Appalachians uplifted with a consequent increase in the downcutting of streams and erosion. The new whale and turtle species recently discovered in the Summerville area by the South Carolina State and Charleston Museums are of Oligocene age. Fauna during this time included large browsing mammals, rodents, cats, birds, horses, alligators, turtles, whales, and dugongs.

The Miocene Epoch

During the Miocene Epoch (22.7 to 5.3 million years ago), South Carolina became colder. The climatic change caused forests to contract and grasslands to spread. In inland parts of North America, small glaciers began to grow. This change, combined with competition between species, resulted in the extinction of 25 percent of all mammal species.

The fauna of the Miocene included whales, dolphins, grazing mammals, early mastodons, dugongs, and seals. The great

Some of the large fossil sharks' teeth (*Carcharadon megalodon*) found in sediments of South Carolina's Coastal Plain

white shark (*Carcharadon megalodon*) developed during this epoch and swam in the sea off South Carolina. This shark grew to 43 feet long, and during its lifetime it could lose over ten thousand teeth—each of which measured over 6 inches long.

Miocene index fossil, *Exogyra*, from the Pee Dee Formation (courtesy: Collection of Carolyn and Jim Smoak)

The Pliocene Epoch

South Carolina remained above sea level for part of the Pliocene Epoch (5.3 to 1.6 million years ago). During the Middle Pliocene, however, there was an inundation by the sea (that created the Orangeburg Scarp). But, sea levels fell again worldwide during the Late Pliocene and a land bridge formed between North and South America at Panama over which animals traveled in both directions. The Pliocene, and the Pleistocene which followed, were characterized by giant mammals. From South America came glyptodonts, giant ground sloths, and armadillos. At the same time, camels, elephants, and horses migrated south. South Carolina fauna also included large carnivores such as cats, wolves, and bears. Grazers included horses, antelopes, and bison. Sea life was abundant.

The Quaternary Period

The final geological period is the Quaternary, which began only a little less than 2 million years ago. Although brief when compared to others, this period has produced a unique burst of animal development. It includes the Pleistocene Epoch (1.6 million to 10,000 years ago) called the "Age of Ice," and the Holocene Epoch (10,000 years ago to the present), known as the "Age of Humans."

R (L) Knobbed Whelk
(*Busycon carica*)

L (R) Horse Conch
(*Pleuroploca sp.*)

From the Pleistocene Wac-camaw Formation in North Myrtle Beach on the Inter-coastal Waterway (courtesy: Collection of Carolyn and Jim Smoak)

The Pleistocene Epoch

During the Pleistocene Epoch (1.6 million to 10,000 years ago), ice covered 30 percent of the earth's surface. Although glaciation did not occur in South Carolina, its animal and plant life, climate, and geomorphology were drastically affected. The average state temperature was 9 degrees (Fahrenheit) cooler in the summer and 32 degrees cooler in the winter than it is now. As the glaciers melted and froze, the seas rose and fell and the climate cooled. Some animals took advantage of the Bering and the Panama land bridges to travel between Asia, North America, and South America. The world's living fauna is a pale shadow of what existed during the Pleistocene. Mammal life then was extremely abundant in North America—comparable to the mod-ern, immense animal-filled grasslands of the Serengeti Plains of Africa with its great herds of grazers and the carnivores that fed on them.

Animals that thrived during the Pleistocene in South Caro-lina included the mastodon, Columbian mammoth, glyptodont, and giant beaver. The mastodon (*Mammut americanum*) originated in Africa and had reached North America by the Miocene. The mastodon, with teeth that had low crowns and rounded cusps, survived by eating crushed leaves, limbs, and tree bark. A warming climate at the end of the Pleistocene Period, about 8,000 years ago, finally brought the mastodon to extinction. But, some researchers credit the Paleo-Indians, who arrived in South Carolina between 12,000 and 14,000 years ago, with helping to speed their extinction through hunting.

Mastodon exhibit at the South Carolina State Museum (courtesy: South Carolina State Museum)

The Columbian mammoth (*Mammuthus columbi*) lived for part of its existence concurrently with the mastodon, and was about the same size. Some mammoths stood over 13 feet high and had tusks 13 feet long. Its teeth were made of flat plates cemented together. Because the mammoth ate coarse vegetation and grasses, it avoided competition with the mastodon. It became extinct about 125,000 years ago, probably due to the Late Pleistocene warming.

The glyptodont (*Glyptotherium floridanum*) stood 3 to 4 feet high and was 8 to 9 feet long. It had a hard, armored shell with small bony plates (called scutes), short legs, and a horned tail. This animal lived in marshes, and ate plants and dug for its food with claws. It was distantly related to the armadillo which today is still thriving in North America. The glyptodont died out during the Late Pleistocene Period, about 6,000 years ago.

Model of glyptodont at the South Carolina State Museum (courtesy: South Carolina State Museum)

The giant beaver (*Castoroides ohioensis*) weighed about 480 pounds (compared to 50 pounds for a modern beaver) and lived in cool lakes and swamps. Its tail was not flat, but was round like that of a weasel. It was the largest rodent in North America during the Pleistocene.

Other important mammals in South Carolina were the giant ground sloths, horses, jaguars, capybaras, tapirs, llamas, camels, dire wolves, bison, rodents, giant land tortoises, bears, giant armadillos, walruses, whales, dugongs, and porpoises. Other animals that often can be found at Pleistocene fossil sites are crocodiles, alligators, sharks, and turtles.

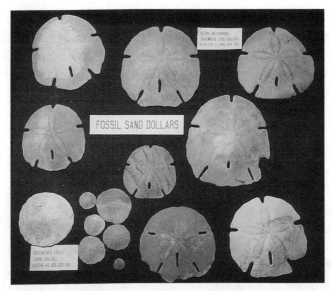

Encope macrophora, arrowhead sand dollar (Pleistocene Epoch) from North Myrtle Beach–Waccamaw Formation; and *Periarchus lyelli,* sand dollar (Eocene Epoch) from the Santee Limestone Formation (courtesy: Collection of Carolyn and Jim Smoak)

Holocene Epoch

Deposits made during the Holocene Epoch (10,000 years ago to the present) are made up of loose sediments that have eroded from higher elevations and moved downslope where they formed deposits in rivers, marshes, bays, and lakes, and along seacoasts. South Carolina's barrier islands and beaches have also formed during the past 10,000 years from reworked sediments. To be considered a fossil, preserved flora and fauna must be older than Holocene.

FOSSIL AREAS OF SOUTH CAROLINA

The Blue Ridge

As discussed in chapter 2, the most recent geological history of the Blue Ridge is one of metamorphism. As a result, most fossil life that might have existed in the rocks before their emplacement during Ordovician time was destroyed by heat and pressure. While parts of the Blue Ridge Mountains in Tennessee and Virginia do contain fossils, none are found in the Blue Ridge area of South Carolina.

The Piedmont

Fossils have been found in the Piedmont in those rocks that were minimally metamorphosed during or after their emplacement. The Piedmont site of greatest historical interest is the Batesburg trilobite metamudstone located in the youngest and least metamorphosed strata of the Carolina slate belt. This find contained over two hundred trilobites that represented nine different species, including the subspecies *Paradoxides davidis grandoculus*. This discovery of 550-million-year-old trilobites, made in 1982 by Sara Samson, a graduate geology student from the University of South Carolina, gave conclusive evidence that the rocks of the Carolina slate belt were not indigenous to North America (Samson 1984).

Polymeroid trilobites of the middle Cambrian age from the upper Asbill Pond Formation of South Carolina found by Sara Samson. (Courtesy: Sara Samson)

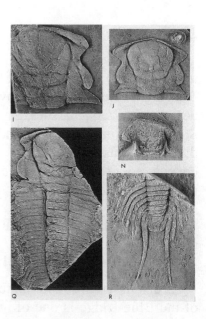

Leaf in clay, Aiken County (courtesy: South Carolina State Museum)

Piedmont petrified wood (courtesy: Read Miner)

Petrified wood and some leaf fossils can be found in the Piedmont where sediments accumulated in river floodplains or swamps long enough to cover and fossilize relatively young plant material. Pieces of petrified wood have been found in Fairfield County and in Laurens County north of the town of Gray Court.

The Coastal Plain

Because of the accessible rocks and sediments and the abundance of life forms that have existed in this region over millions of years, the Coastal Plain contains the vast majority of South Carolina's fossils.

Explore, either on your own or with experienced fossil club members, areas where underlying rocks and sediments are exposed in the Coastal Plain. Likely fossil areas include ditches; river banks; beaches; gravel, sand, and lime quarries; and anywhere that industry, construction, or forestry companies are grading or otherwise digging into the earth. Be sure to have any necessary permission and permits before beginning your exploration.

Listed below are formations in the region that contain fossils and a number of sites where exploration is recommended. Geologic maps can be purchased from the South Carolina Geological Survey Office (see the resources list for the address) to help you find new sites to explore.

Coastal Plain Fossils

The rocks and sediments of South Carolina are divided into formations, which are rock layers that have distinctive lithic features in common, and so are represented as a unit. The rock formations of the Coastal Plain are not found continuously all over the province because of variances in depositional and erosional histories (see chapter 5). While the Orangeburg Scarp is accepted as the boundary between the upper and middle Coastal Plain, some of these formations extend beyond the escarpment. (For a detailed description of these formations, refer to Horton and Zullo, *The Geology of the Carolinas*.)

The table below outlines formations that contain macro, or large, fossils:

Table 8.1   Macro Fossil-Bearing Formations of South Carolina: Upper, Middle, and Lower Coastal Plain

Period	*Epoch*	*Formation*
Quaternary	Pleistocene	Talbot
		Waccamaw
Tertiary	Pliocene	Duplin Marl
	Miocene	Hawthorne
	Oligocene	Chandler Bridge
		Cooper Group:
		Ashley
	Eocene	**Barnwell Group:**
		Dry Branch
		Tobacco Road
		Cooper Group:
		Parkers Ferry
		Harleyville
		(Upper) Cross
		Orangeburg Group:
		Lower Cross
		McBean
		Huber
		Santee Limestone
		Congaree
	Paleocene	**Black Mingo Group:**
		Lang Syne
Cretaceous		**Black Creek Group:**
		Pee Dee
		Donoho Creek

The Upper Coastal Plain

The fossiliferous formations in the upper Coastal Plain range from Paleocene (Lang Syne Formation), to Eocene (the Orangeburg Group and the Barnwell Group).

Black Mingo Group

The Black Mingo Group includes the Lang Syne Formation. This formation contains pebbles, sands, thin beds of black clay, and pockets of fullers earth. Fossils found within it include sharks' teeth, annelids, dinoflagellates, corals, mollusks, worm burrows, nannofossils, and pollen. Lang Syne is well exposed along the shore of Lake Marion, in western Sumter County along the Wateree River, in Lee County at Red Hill, in Calhoun County along the Congaree River, and on the grounds of Fort Jackson in Richland County.

Orangeburg Group

The Huber Formation contains fossil leaf imprints in clay beds in Aiken County. Exposures can be found along the major tributaries flowing into the Savannah River: Little Horse Creek, Horse Creek, Town Creek, and Hollow Creek. The exposures can also be found along the south fork of the Edisto River running from outside the town of Johnston in Edgefield County to below Aiken State Park. The Congaree Formation can be found on the hilltops throughout the High Hills of Santee in western Sumter and southwestern Lee Counties, and along the Congaree River and its tributaries. The Congaree Formation is only lightly fossiliferous, but clams and burrows have been found in it. The McBean Formation is made up of silicified coquina, siltstones, and sandy and clayey marls that contain over sixty-five species of marine fossils. Exposures are numerous and include the area near St. Matthews in Calhoun County, along Tinker Creek and Upper Three Runs Stream in Aiken County, and along the North Edisto River from the city of Orangeburg north to Park Crossroads in Orangeburg County.

Barnwell Group

This group contains fossiliferous Eocene formations, which include sandstones, sand, and clays. *Turritella* and bryozoans are found in cherts along the west bank of the Savannah River south of Augusta. Fossil oysters, crinoids, and barnacles from the Dry Branch Formation have been found at Usserys Bluff in Allendale County.

Middle and Lower Coastal Plain

The middle and lower Coastal Plain have many more fossil sites than the upper Coastal Plain. The fossil-bearing formations range in age from Cretaceous (Donoho Creek and Pee Dee), to Paleocene (Black Mingo), to Eocene (Orangeburg and Cooper Groups), to Oligocene (Cooper), to Miocene (Hawthorne), to Pliocene (Duplin), and finally to Pleistocene (Waccamaw and Talbot).

Listed below are a few fossil sites in the middle and lower Coastal Plain.

The Black Creek Group

The Donoho Creek Formation

The Black Creek Group is Late Cretaceous and includes the Pee Dee and Donoho Creek Formations. The Donoho Creek Formation contains tidal mud flat sediments. It is made up of dark and buff sands with worm and crab burrows. The formation lies along the Great Pee Dee River in Florence and Darlington Counties. It contains marine fossils including belemnites, *Exogyra,* and other mollusks. The South Carolina dinosaur discoveries of the past few years have come from this formation.

The Pee Dee Formation

This formation is the youngest of the Cretaceous, and is found in Horry, Georgetown, Florence, and Williamsburg Counties. It is made up of dark gray, sandy marls and black clays. Several types of oysters, belemnites, mollusks, worm burrows, reptiles, mosasaurs, plesiosaurs, and whales are found in this formation. The important index fossil, *Exogyra costata,* is found in the Pee Dee Formation.

Belemnites from the Great
Pee Dee River at Burches
Ferry, Florence County
(courtesy: Read Miner)

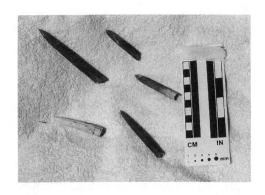

There are many good fossil sites in Florence County along the bluffs on the west bank of the Great Pee Dee River. These sites include Burches Ferry, Mars Bluff, Orum, Dewitts Bluff, and Cains Landing—all near Pamplico.

Black Mingo Group

This group includes Paleocene sands and clays that contain a small group of fossils. The Lang Syne Formation is found on the shores of Lake Marion and in stream systems below the Orangeburg Scarp.

Orangeburg Group

The Santee Limestone Formation is a creamy yellow to white limestone that is found on the shores of Lake Marion, and in Dorchester and Orangeburg Counties. It contains bryozoa, echinoderms, and mollusks (including oysters). It also contains sharks' teeth, nautiloids, sawfish, and leatherback sea turtles. A popular fossil hunting site is the Giant Portland Cement Company Quarry near Harlcyville in Dorchester County.

The Cross Formation is well exposed at quarries in Berkeley, Orangeburg and Dorchester Counties. Fossils include many marine species.

Cooper Group

These Eocene and Oligocene formations are well known as wonderful sites for collecting marine fossils such as whales and sea cows, sharks' teeth, turtles, and alligators. It also contains a rich array of corals, bryozoans, echinoderms, brachiopods, gastropods, annelids, cephalopods, and crustaceans. Notable collecting sites include the Giant Portland Cement Company—the site of the discovery of the proto-whale, archeocete (*Dorudon serratus*), at the base of the Cooper Marl.

Chandler Bridge

Sites in and around Summerville have recently been excavated by the Charleston and South Carolina State Museums. These excavations have led to the discovery of new species, such as odontocete whales. This appears to be a relatively untapped and very rich area for fossils. Recently, an uncataloged rodent species was found by an amateur paleontologist in a backyard site.

The Hawthorne Formation

A sparsely fossiliferous Miocene formation, the Hawthorne, is found along the Savannah River. This formation is made up of brittle clays and a phosphatic marl.

Duplin Marl

The Duplin Marl is a three-to-five-million-year-old Pliocene formation that contains over three hundred species of marine mollusks and foraminifera—some of which are found in shell marl. Tearcoat Branch Creek in Sumter County is the site from which many of the fossils, including many varieties of mollusks now at the McKissick Museum in Columbia, were found.

The Waccamaw Formation

The Waccamaw Formation is a 1.6-to-2.5-million-year-old (Pleistocene) formation, which, in Horry County, overlies the Cretaceous Pee Dee Formation. These diverse formations create an unconformity that spans 62 million years. The proximity of the two formations to each other sometimes allows for the collection of fossils from both ages in one location. Over five hundred species of mollusk have been found in the Waccamaw, which makes it among the top fossil collecting sites in the state. Fossils from this formation include many species of gastropods, pelecypods, corals, bryozoans, echinoderms, and sharks' teeth. These fossils occur in the low tide sands all along the Intracoastal Waterway, but the best sites are north of U.S. Highway 501.

The Talbot Formation

This two-hundred-thousand-year-old Pleistocene formation includes within it mostly modern marine species, with only 2 percent extinct species. The Talbot represents an old barrier island, estuarine, and shelf floor environment. It is made up mostly of clays, sands, and gravels in which mollusks, corals, and sharks' teeth are found. Many exposures in Horry County and at Yonges Island near Charleston are from sediments in the Talbot Formation. The shoreline of the Intracoastal Waterway in Myrtle Beach is a favorite spot for fossil hunters.

The Ashley River in Dorchester County, about 9 miles north of Charleston, has been a well-worked fossil site for over 100 years. The deposit consists of thin muds that hold the fossils. Those muds are overlain by clay-banded sands. Included in the deposit are horses, tapirs, lions, pumas, bears, saber-toothed tigers, mammoths, and mastodons. These faunas are considered to be Late Pleistocene, from about 40,000 to 100,000 years old.

Several sites in the state along the great rivers have produced fine petrified wood. Abundant finds are possible by looking along the floodplains of the Pee Dee, the Wateree, and the Pocotaligo Rivers near gravel pits. Sites near towns such as Boykin in Lee County, Blenheim in Marlboro County, and Manning in Clarendon County have produced good specimens. Other counties with abundant fossil wood include Sumter, Colleton, Clarendon, Florence, Darlington, and Dillon. Fossil wood includes palm, maple, cypress, and sweet gum.

Underwater collecting can be done with adequate training and a license obtained through the South Carolina Institute of Archaeology and Anthropology (see resources list). Permission should always be obtained before hunting fossils on private property. All sites must be treated with respect and left in the conditions in which they were found.

Fossil Displays in South Carolina

The following sites have fossil collections that are open to the public:

1. South Carolina State Museum, Columbia
2. McKissick Museum at the University of South Carolina, Columbia
3. Charleston Museum, Charleston
4. Charleston Southern University, Charleston
5. Edisto Beach State Park, Edisto Island
6. Clemson University, Clemson
7. Furman University, Greenville
8. Museum of York County, Rock Hill
9. Spartanburg Science Center, Spartanburg

Interior of room at the O'Donnell residence, 21 King Street, Charleston (credit: George Cook)

CHAPTER 9

THE CHARLESTON EARTHQUAKE

On August 31, 1886, the Charleston area experienced a major earthquake—the most destructive ever recorded in the eastern half of North America. The New Madrid, Missouri, earthquake of 1811 was more intense than both the Charleston earthquake and the San Francisco earthquake of 1906, but comparatively little damage to structures occurred there because of the small population living in the region at the time. In contrast, San Francisco and Charleston sustained major damage. All of these areas are designated as potentially dangerous seismic zones.

Modern Charleston street scene

The Charleston earthquake consisted of two main shocks: the first at 9:51 in the evening, and another eight minutes later. Several foreshocks had been felt in the area more than ten days prior to the evening of August 31. Ironically, the August 31 edition of the Charleston newspaper, the *News and Courier,* included accounts by several people who reported that the earth moved toward the southwest on and before August 21 in the Oakley area and along the Cooper River.

Information from diaries, letters, and newspaper articles from the period suggests that the earthquake would have measured (using modern scales) a force of X on the Modified Mercalli Scale, a moment magnitude of 7.7, and a 7.6 on the Richter Scale. Because no such scales existed in 1886, geologists have

Twisted railroad tracks outside of Charleston (credit: George Cook)

Tradd Street in Charleston (credit: George Cook)

St. Phillips Church—damaged in the earthquake
(credit: George Cook)

had to piece together the details of the earthquake, including its
intensity and location, by using geological and geophysical evidence, eyewitness accounts, and photographs. Geologists have
estimated that the epicenter of the earthquake was located about
15 miles northwest of Charleston near the present site of
Middleton Place, on the Ashley River. The rupture is thought
to have been 18.6 miles long and to have occurred at a depth of
7.4 miles, with an average slip (or displacement) of about 5.6 feet.

The Ashley River just north of Middleton Place. This area is
thought to have been near the epicenter of the Charleston earthquake.

Ninety percent of the brick structures in Charleston were damaged during the earthquake. Buildings cracked and split, and most of the 110 people who lost their lives were killed by falling debris. Over fourteen thousand brick chimneys collapsed. Railroad tracks were thrown out of alignment, telegraph lines snapped, and many small fires erupted throughout the city. It was reported that some citizens smelled sulphur gas coming out of the marshes shortly after the quake, and field hands saw a small tidal wave bore up the Cooper River. Many people ran out into the town squares and camped there for several nights rather than risk returning to their homes. This action may well have prevented further deaths from falling rubble, as aftershocks continued every few hours for several days.

Wooden structures fared far better than those made of masonry or brick because the wood bent and gave at the joints and then sprang back into its original shape. The greatest damage to the city was located in the eastern and southern sections which included the business district, the wharves, and the Battery. While 90 per-

Residents camped in town squares, like Washington Park, after the earthquake, fearing returning to their homes (credit: George Cook)

cent of the brick buildings suffered damage, only 7 percent of the frame buildings were hurt—mainly from falling chimneys, cracked plaster, and damaged foundations. Buildings set on solid ground fared far better than those built on the filled land that comprised a large portion of Charleston in 1886. The city also suffered damage to its cemeteries, as gravestones tipped over and fell in all directions. The fallen positions of the gravestones provided investigators with valuable evidence that during the quake the movement of the earth had been both horizontal and vertical.

Damage was reported in all of the large towns situated within 200 miles of Charleston. In Savannah, three hundred chimneys were damaged and some buildings cracked. In Columbia, many residents reported that they had experienced difficulty walking upright during the time the earthquake occurred and that buildings swayed and plaster fell. To prevent future damage, some buildings were reinforced with earthquake rods. Shocks were felt within an eight-hundred-mile radius, from Cuba to New York and from Bermuda to the Mississippi. Because the geology of the subsurface largely determines where an earthquake is felt, cities such as Columbia (which sits on top of dense sands and granitic bedrock of the Fall Zone) resonated like a tuning fork as the seismic waves hit. The earth slumped in many places

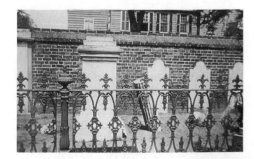

Displaced monuments at St. John's Lutheran Church, Charleston (credit: George Cook)

McCutchen House on the Horseshoe at the University of South Carolina, Columbia, showing earthquake rods.

and the sandblows, which are still visible today, are evidence that the Coastal Plain also shook violently during the earthquake.

Communications into Charleston were restored in a matter of a few days after the earthquake. The people who remained in the city (some took the first working train out) immediately began the process of repairing and rebuilding. Help came from all over as individuals, businesses, and government officials offered funds and materials. The following are some of the telegrams and telegraphs received and printed in the *News and Courier* on September 4, 1886:

> Saratoga, N.Y.: Have just heard details of your calamity. Should assistance be needed, please draw on the Bank of New Hanover of Wilmington, N.C. against my account for $250.
>
> William Latimer, Wilmington, N.C.

> Richmond, Va.: Richmond Medical and Surgical Society tenders sympathies and professional services if any doctors are needed. Telegraph us how many. We will bear expenses.
>
> John G. S. Kelton, M.D., President.

> [The governor of North Carolina sent the following dispatch to Governor Sheppard of South Carolina:] We have news of terrible calamities in your state. How can we best aid your people? Will gladly come to their relief.
>
> A. M. Scales.

Even the queen of England was moved by the catastrophic events in Charleston and quickly sent the following message to President Grover Cleveland, as printed in the *Columbia Daily Register* on September 4, 1886:

Balmoral, September 3, 1886

To the President of the United States:

I desire to express my profound sympathy with the sufferers by the late earthquakes, and await with anxiety fuller intelligence, which I hope may show the effects to have been less disastrous than reported.

Victoria Regina

The Queen's Cablegram (*City of Charleston: The Yearbook of 1886*)

The total damage costs of the 1886 earthquake (in 1886 dollars) have been estimated at $5.5 million dollars. Scientists have collected geological data suggesting that Charleston would not escape great financial hardship if another earthquake measuring 7.6 on the Richter Scale were to hit the area today. Of special concern to both geologists and local governments is the knowledge that if another earthquake of this size were to occur, damage of every kind would surely be catastrophic—affecting millions of people in South Carolina and throughout the southeast. While an earthquake of such magnitude will probably not hit the Charleston area in the near future, earthquakes of lesser magnitude (6 or less on the Richter Scale) are a stronger possibility. The damage caused by future earthquakes would still be tremendously harmful to the area. With the state's infrastructure at risk (which now includes nuclear power plants, highways, dams, and military installations, as well as homes and businesses), scientists are studying the geology of South Carolina with some urgency to determine the causes of the seismicity of the area and thereby calculate future risk.

Was the event of 1886 unique? What is the seismic rate in the Charleston area for great earthquakes? Could earthquakes strike other areas along the eastern coast? Answers to these questions have today been partially determined. Since the early 1970s, geophysical studies have been carried out in the Charleston area, along the adjacent Coastal Plain, and offshore to the edge of the continental shelf. While experts are not yet in complete agreement on either the geological structures underlying Charleston or on the forces at work under the earth that gave rise to the great earthquake, much has been discovered.

Because of the geology of the Charleston area, with its huge overburden of sediments, no fault line can be seen at the surface from either the 1886 earthquake or from any others. Therefore, much research has centered on surface and near-surface evidence for ancient, as well as more recent, earthquakes. Geologists who study these ancient earthquakes are called paleoseismologists. Evidence of earthquakes that occurred as far back as 7,200 years ago can be seen in the effects of the liquefaction and ejection of soils around Charleston and Summerville in craterlike structures called sandblows. Sandblows are geological "fingerprints" that are created within the pore spaces in the soil during an earthquake as the soil particles transfer their weight to the water molecules that lie between them. Because the water is restricted, the pore water pressure increases to equal that of the weight of the soil, causing the soil to behave briefly like

a liquid. This soil shoots out of the ground, leaving characteristic sandblow craters in the epicentral area. In Charleston, these craters range in size from a few inches to 20 feet
in diameter. Other sandblows are located throughout the
Coastal Plain and vary greatly in age and size. The evidence
provided by the soils of the sandblows, and by the organic
material found within them, gives valuable evidence of
earthquake age. Plant roots embedded in the soil are radiocarbon dated, telling geologists the approximate time in
which the earthquakes that caused the sandblows occurred.
Sandblows, therefore, are being intensely studied today in
South Carolina because they give researchers some of the
best evidence about the dates and directions of movement
along Charleston's hidden faults. Sandblows can also be
found in the Georgetown area and farther north along the
coast into North Carolina.

The technique of seismic reflection has been used in the
Southeast to further investigate Coastal Plain seismicity and
aid in the continuing national search for oil. A group called the
Consortium for Continental Reflection Profiling (COCORP),
made up of university and oil company scientists, has intensi-
fied these studies over the past 10 years. By sending seismic
waves into the earth, scientists can determine the relative

Sandblow in Summerville, South Carolina, 1886
(credit: George Cook)

densities of the rock by how the waves differentially bounce off the underground strata. The waves reflect up to instruments at the surface where they are recorded and interpreted, providing an outline of the underground geology. The pictures made using COCORP techniques have allowed geologists to produce a sort of relief map of the land under South Carolina.

Seismic waves continue to be carefully monitored in the Coastal Plain as well as in the Piedmont and Blue Ridge. Seismometers have been stationed at several localities around the state since the 1930s. These instruments, which pick up the slightest earth movement, have recorded hundreds of local and regional earthquakes, as well as distant major ones, which have occurred during the past 60 years.

Gravity and magnetic anomaly studies have been made in an attempt to read the nature of the geology far underground. These studies use instruments that are flown over the test area in order to measure the gravity and magnetic readings of the rocks below. These data show what general kinds of rocks lie below the surface. With this information, geologists can often determine the location of faults.

Charleston and the rest of eastern North America are now part of a passive continental margin where no mountain-building activity has occurred for over 180 million years, but the fault scars of the continent's past tectonics exist underground. While there are several faults in the Charleston area, the Woodstock Fault and the Ashley River Fault are thought to be the two associated with Charleston's 1886 earthquake. The following faults are found in the area:

1. *Woodstock Fault*—thought to be a right-lateral strike-slip fault 5.5 to 8 miles deep that trends north 26 degrees east for about 25 miles from Jedburg to Adams Run.

2. *Ashley River Fault*—a southwest-dipping reverse fault 2.5 to 4.5 miles deep that trends north 37

degrees west for over 3 miles on the Ashley River near Summerville.

3. *Cooke Fault*—a reverse fault that trends north 70 degrees east for 6 miles between the Ashley and Edisto Rivers just south of Summerville, thought to be inactive.

4. There is evidence for a deep, steep-angled fault that is part of a graben system of Mesozoic rifts at about 7.4 miles below the surface.

5. There is evidence for a low-angle thrust fault beneath the entire region.

Dr. Pradeep Talwani, of the University of South Carolina, Geology Department, has proposed that the Charleston earthquake was caused by the movement of the Woodstock Fault and that the aftershocks can be attributed to movements of the shallower Ashley River Fault. Because the two faults cross each other and lock together directly under the Summerville area, this part of South Carolina is designated a permanent zone of stress and can expect destructive earthquakes in the future.

Many geologists feel that Charleston's seismicity is related to its position on top of an ancient rift zone, which was formed as the North American plate pulled away from the African plate beginning in the Late Triassic. The upper crust thinned as the two plates rifted apart, much like chewing gum or silly putty thins out as it stretches. As the plates pulled apart, pieces of crust dropped to form grabens. Over time, sediments filled and covered these grabens and today they lie about 7.5 miles under the surface of the Coastal Plain.

Charleston also sits at the northeast corner of the warp that was caused by the formation of the Cape Fear Arch, the Charleston Embayment, the Yamacraw Arch, and the Southeast

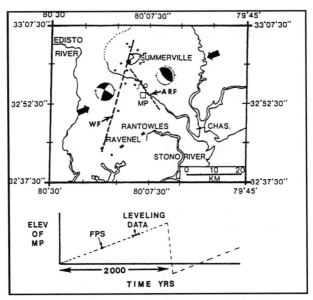

Location of faults under the Charleston area: Woodstock and Ashley River Faults (courtesy: Dr. Pradeep Talwani)

Georgia Embayment. The Charleston Seismic Zone has been active, then, at least since the Eocene (48 million years ago).

Many geologists also feel that the compression that caused the movement along the faults under Charleston, which culminated in the Charleston earthquake, is the prevailing east–northeast/west–southwest to northeast/southwest pressure common to the central, northeastern, and southeastern portions of the North American plate at this time. Fault movement in the eastern United States has increased a great deal since the Late Cretaceous (ranging from 1.1 to 5 feet per million years). Because Charleston and the surrounding areas lie on a compressed and rifted zone (as shown by COCORP readings, gravity anomalies, and well drillings), the land has a severely fractured subsurface. This makes it appear likely that the underground faults left by the collision and retreat of the North American and African plates, and the constant and generalized continental pressure, may well be the ultimate causes of Charleston's seismicity.

Geologists believe that over the past 5,000 years the Charleston area has been hit by as many as five large earthquakes and that a large earthquake preceded the 1886 event by 1,100 years.

Table 9.1   Earthquake Recurrence Rates in the Southeast

Modified Mercalli Scale	Richter Scale	South Carolina	Southeast
VI	5	10 years	3 years
VIII	6	100 years	20 years
X	7	Unknown	Unknown

Return Periods for a Typical Site in South Carolina

Modified Mercalli Scale	Return Period at any Given Site
VI	400 years
VIII	10,000 years
X	Unknown

Given the nature of urban congestion, it is probably unrealistic to assume that modern American cities could ever really be prepared for a major earthquake. The devastating earthquakes of this century (Armenia, Peru, San Francisco, and Iran in 1989 and 1990; Los Angeles in 1994) testify to the inherent unpreparedness and inadequacy of human-made structures and systems when confronted with the determined, moving earth.

Many steps can be taken, however, to lessen the damage and loss of life. Individuals can learn the safest places to be during an earthquake, both inside and outside of buildings; they can have on hand at all times emergency food and medical supplies; and they can create safer surroundings by making their own dwellings as earthquake resistant as possible. Collectively, the people in South Carolina's cities and towns can make many important changes to existing buildings and can insist on stricter building codes for new construction in earthquake-prone areas.

Weak buildings can be taken down, dangerous architectural ornaments removed, and highway overpasses reinforced. Government agencies can also upgrade their emergency preparedness to include the improved and accelerated education of all citizens using the media, churches and schools, and local and national organizations.

The federal government has helped South Carolina take some important steps to achieve the goals of earthquake preparedness by establishing, in 1976, the Seismic Monitoring Network and Earthquake Education Center at Charleston Southern University in Charleston. Today, seven stations located near the 1886 epicenter feed their seismic signals directly to the Center, which then sends the data to the United States Geological Survey offices in Colorado. The seven stations, which listen to the ground twenty-four hours a day, are set up at James Island, Hollywood, Middleton Place, S.C. Highway 61 (near Middleton Place), Tillmans/Whites Bays (near St. George), U.S. Highway 17 (Charleston, west of the Ashley River), and Charleston Southern University. The Piedmont region is also closely watched with instruments monitored by the University of South Carolina. At least 351 earthquakes were recorded from 1886 to 1913 in the Charleston area. From 1973 to 1993, 190 earthquakes were recorded in South Carolina. The largest earthquake recorded in the state during 1973 to 1993, which occurred on August 21, 1992, in the Charleston/Summerville area, measured a magnitude of 4.1 on the Richter Scale. All others ranged from less than 1 to 3.8, with most by far located along or near the Middleton Place Seismic Zone.

The Earthquake Education Center hopes that this constant monitoring of the earth under Charleston will provide adequate warning of any increase in seismic activity, and that it will give scientists, government officials, and individuals enough time to better prepare for a major earthquake. The Center educates the public in earthquake preparedness through an outreach program to schools and community organizations, and by its publication of written materials for the public (see resources list).

Seismometer at the Seismic Monitoring Network and Earthquake Education Center at Charleston Southern University in Charleston

The ongoing seismic research, which is presently in progress by both public universities and private businesses, reminds us that there is still a great deal to be discovered about the formation of the earth, and about the tectonics and seismicity of the southeastern region. All who live in South Carolina should make it their business to become as informed as possible about earthquakes and how best to survive them.

PROTECTION OF SOUTH CAROLINA'S NATURAL RESOURCES

The varied geographic regions of South Carolina were created out of an ancient geological and climatic history. Today, development in the state is largely dependent upon modern human influences. The human-made alterations of the natural landscape have created economic growth, but many of these alterations have created problems that pose great environmental and ecological challenges. Some of these challenges include the need to protect groundwater, clean up stream and lake pollution, find wiser means of disposing of toxic wastes, counter the loss of habitat for animals and plants, replace beach sands lost through the damming of rivers, plan and limit urbanization and development, and protect and replace the topsoils lost through natural erosion and poor farming practices.

Although much of South Carolina's lands are threatened by overdevelopment, urbanization, erosion, pollution, and unwise management, there remain thousands of acres of unspoiled beauty to study and enjoy. Fortunately, today many people of South Carolina—business, government, landowners, and private citizens—are engaged in a concerted effort to find solutions that will best maintain the natural health of the state. Time is short, however. Only if we all aspire to become better caretakers of our natural heritage, rather than merely its exploiters, will these rare and delicate environments remain with us. It is in the hope of fostering future trustees of the land through knowledge and awareness that this book is written.

APPENDIX

APPENDIX A

South Carolina Radon Levels

Granitic rocks and phosphates contain trace amounts of radioactive elements, such as uranium, that decays ultimately into radium, and then into radon and other daughter elements. The highest radon gas readings in South Carolina are found, not surprisingly, in those upcountry counties underlaid by granite and gneiss. According to the Environmental Protection Agency (EPA), radon gas is the second greatest cause of lung cancer in the United States (smoking being the first). The EPA also notes that it is especially important for residents in the upcountry to check their homes for the presence of radon gas. In South Carolina, igneous intrusive rocks and metamorphic gneisses, in particular, can pose a health danger if human-made structures built on top of them are allowed to trap the radioactive gases. The South Carolina Department of Health and Environmental Control has noted that measurements of radon gas greater than 4.0 picocuries per liter should promote immediate remedial actions in a building so affected. South Carolina has an average radon level of only 1.1 picocuries per liter, with higher than average levels found in some areas of the Piedmont and Blue Ridge regions.

APPENDIX B

South Carolina Gemstones of the
Blue Ridge and Piedmont

Numerous gemstones have been discovered in South Carolina. Below is a listing of some of them and the counties in the Blue Ridge and Piedmont regions where they can be found.

Gemstone	County
Amethyst	Abbeville, Anderson, Cherokee, Greenwood, Union
Aquamarine	Anderson, Cherokee
Beryl	
Goshenite	Anderson, Cherokee
Heliodor	Anderson, Cherokee
Citrine Quartz	Fairfield, Kershaw
Chalcedony	Allendale, Chesterfield
Clear Quartz	Abbeville, Anderson, Cherokee, Fairfield, Greenville, Greenwood, Kershaw, Richland
Corundum	Cherokee
Diamond	Spartanburg
Emerald	Anderson, Cherokee
Epidote	Lexington, Union
Garnet	Anderson, Greenville, Lexington, Pickens
Gold	Cherokee, Chesterfield, Fairfield, Kershaw, McCormick, Union, York
Hematite	Newberry
Jasper	Many areas, mainly northern and western counties
Kyanite	Lexington, Newberry
Lace Agate	Aiken
Lazulite	Lexington, Newberry
Moonstone	Anderson, Kershaw, Lexington, Union
Olivine (Dunite)	Union
Petrified Wood	Fairfield, Laurens

Gemstone	County
Ruby	Anderson, Cherokee, Laurens, York
Rutile	Lexington, Newberry
Sapphire	Anderson, Cherokee, Laurens, York
Spodumene	Cherokee, York
Smoky Quartz	Abbeville, Anderson, Cherokee, Fairfield, Greenville, Greenwood, Kershaw
Sillimanite	Cherokee, Oconee
Topaz	Anderson, Cherokee, Chesterfield
Tourmaline	Anderson, Greenville, Spartanburg
Unakite	Lexington, Union, York
Zircon	Anderson, Cherokee, Greenville

GLOSSARY

abyssal plain—The flat, almost level area of the ocean floor.

accretion—The gradual addition of new land to old.

Agnatha—A class of primitive vertebrates, the jawless fish, that includes the lampreys.

allochthonous—Materials formed elsewhere than in their present place. Of foreign origin.

alloy—A mixture of metals.

ammonite—Extinct shelled marine animals that lived from the Devonian to the Cretaceous.

amphibolite—A metamorphic rock consisting mainly of the minerals amphibole and plagioclase, with little or no quartz.

anaerobic—Pertaining to or caused by the lack of oxygen. Of an organism living without oxygen.

analogue—A partial similarity between like features of two things, on which a comparison may be based.

annelid—A worm phylum characterized by a segmented body. Fossil remains, usually burrows or trails.

anomaly—A geological feature, usually in the subsurface, which is different from the general surroundings.

anticline—A fold of rock, usually convex upward like an arch.

aquiclude—A nonporous formation, such as shale, that slows or stops the passage of water underground.

aquifer—A porous zone of underground rock or sediment, such as sandstone, in which water is found.

arch—An anticline.

archeocete—One of the three orders of whales.

asthenosphere—The earth's hot, plastic upper mantle.

Avalon terrain—The exotic microcontinent that attached to North America during the Paleozoic and formed the Carolina terrain in South Carolina.

backshore—A zone extending inland from a berm to the farthest point reached by waves.

Baltica—The early continent that was to become Europe. Proto-Europe.

barrier island—Sand deposits that parallel a shore but are separated from the mainland.

basement rock—Rock, commonly igneous and metamorphic, that underlies the rocks of interest (which are often sedimentary).

basin—A low area in the earth's crust of tectonic origin in which sediments have accumulated.

batholith—A body of intrusive igneous rock, such as granite, that is at least 40 square miles in area.

bay—A wide, open curving indentation or inlet of a sea or lake onto an adjacent land mass.

beach ridge—A continuous mound of beach material, commonly sand, which has been heaped up by wave and wind action.

bedrock—Solid rock found under soil.

belemnite—An extinct type of cephalopod, recognized by cigar-shaped fossils of parts of its internal skeleton.

berm—A nearly horizontal portion of the beach or backshore formed by the deposit of material by wave action.

bivalve—A mollusk of the class Pelecypoda having two distinct shells.

black river—A river that carries little sediment and is colored dark by tannins formed from decomposing vegetation.

black smokers—Undersea hot water vents that deposit minerals on the sea floor.

blastoid—A class of stemmed, budlike echinoderms having fivefold symmetry.

bog—Waterlogged, spongy ground consisting primarily of mosses containing acidic, decaying vegetation that may develop into peat.

brachiopod—Any marine invertebrate belonging to the phylum Brachiopoda characterized as having two shells.

breaker line—The most landward line along a shore at which waves hit.

breccia—A coarse-grained clastic rock composed of angular broken rock fragments held together by a mineral cement.

bryozoans—Any marine invertebrate belonging to the phylum Bryozoa and characterized by colonial growth and a branching, twiglike skeleton.

Burgess shale—A shale from the Canadian Rockies that contains rare Cambrian fossils of soft-bodied marine life.

capybara—A large rodent common during the Pleistocene.

carbonate banks—Thick layers of limestone or dolomite formed in warm, shallow seas.

carbonization—The slow decay under water of organic material that results in a concentration of carbon as a film showing, more or less distinctly, the form and structure of the original material.

Carolina bay—A generally oval-shaped depression found on the eastern Coastal Plain thought to be caused by wind and tidal forces.

cartilaginous—Having a nonbony skeleton made from cartilage, such as that of a shark.

cast—A method of fossil preservation that occurs when a mineral or sediment fills a fossil mold.

cephalopod—A marine mollusk of the class Cephalopoda, characterized by a head surrounded by tentacles and, in most fossil forms, by a calcareous shell divided into chambers.

chemical differentiation—The process by which minerals leave a melt as they solidify due to their different melting points. As minerals solidify out of a melt, the potential for mineralization is altered for the liquid minerals that remain.

chert—Nonclastic sedimentary rock made of silica.

Chordata—A phylum of vertebrates having a notochord (or spinal cord) usually protected by a backbone.

clastic—Broken. Rocks or sediment that are formed from broken pieces of former rocks or shells. Examples include sandstone and shale.

climax forest—The final stable or equilibrium stage of development that a flora goes through in a given environment. A mature community.

COCORP—Consortium for Continental Reflection Profiling. A group of oil companies and university scientists who have organized to map the area beneath the earth's surface using seismic reflection.

compaction—process by which sediments are pressed over time to become rock.

contact—The boundary between two types or ages of rocks.

continental crust—The rock underlying the continents and continental shelves that is composed mainly of granite and granodeorite and contain large amounts of silica-rich and aluminum-rich minerals.

continental drift—The movement of continental plates thought to be caused by convection in the mantle.

continental margin—The various areas between the shoreline and the abyssal plain, including the continental shelf, continental slope, and the continental rise.

continental shelf—The relatively flat part of a continent that lies between the shoreline and the continental slope covered by seawater.

convection current—The flow of heat in a fluid due to density differences.

convergent plate boundary—The boundary at which two crustal plates collide with each other.

coquina—A limestone composed of broken, abraded shell fragments.

country rock—The rock surrounding a mineral deposit or an igneous intrusion.

craton—The eroded core and oldest part of a continent.

crenulations—Small folds with a wavelength of a few millimeters, or fractions of inches, chiefly found in metamorphic rocks.

crinoid—A marine echinoderm characterized by a globular body enclosed in a calyx from which arms extend radially with a flexible stem and attachment to the sea floor.

Crossopterygian—An extinct group of fish, except for the coelacanth, regarded as being ancestral to the amphibians and other land vertebrates.

crust—The thin, outermost layer of the earth that lies atop the mantle.

crustacean—An arthropod, such as a shrimp, characterized by two pair of antennae.

crustal rebound—The rise of continental rock as weight is removed to achieve isostasy. For example, after the glaciers melted the land rose or rebounded.

crystalline rocks—Rock formed from minerals having regular molecular structure. Igneous or metamorphic rocks, rather than sedimentary rocks.

cyanide-leaching process—In gold refining, the process used to chemically extract gold from ore using chemicals such as cyanide.

deciduous—Trees or shrubs that shed their leaves annually.

delta—A fan-shaped deposit of sediments that accumulates when a moving body of water loses its velocity.

dendritic—A treelike, branched pattern. A river or stream system that has many tributaries that flow into a central stream.

diabase—A rock of basaltic composition, consisting essentially of the minerals labradorite and pyroxene, that commonly forms dikes and sills.

diatom—A microscopic, single-celled aquatic plant related to the algae.

dike—A tabular body of molten rock that intrudes, cutting across the structure of adjacent rocks.

dike swarm—A group of dikes that flow in radial, parallel, or steplike arrangement.

dinoflagellates—A one-celled, mostly marine, organism having a flagellum that resembles aspects of both the animal and plant kingdoms.

divergent plate boundary—The boundary at which two crustal plates move away from each other.

domes—An uplift or anticlinal structure in which the rocks dip gently away in all directions.

dore—An ingot of gold or silver.

downdrift—In the direction of the longshore current. Downstream.

downwarp—An area that has been bent downwards, usually in a broad syncline.

dromeosaurs—A large group of carnivorous dinosaurs.

dugong—A herbivorous, aquatic mammal having a fishlike body, flipperlike forelimbs, no hind limbs, and a paddlelike tail. Common fossils found in the Coastal Plain.

ebb tide—That part of a tide cycle between high water and the following low water characterized by a seaward or receding movement of water.

ebbtide delta—Sandy shoals found offshore at the mouth of an inlet created by the flow of ebbtide waters moving seaward.

echinoderm—Any species of the phylum Echinodermata, such as the starfish. Marine benthic invertebrates characterized by radial symmetry. Mostly consisting of rigidly plated bodies and a water-vascular system.

ecosystem—A community of different species interacting with one another and with the chemical and physical factors making up their environment.

eddy—A current of water that runs counter to the main current, especially one moving in a circle. A whirlpool.

ejecta breccia—Rock pieces broken and heated by meteorite impact and fused into a new rock.

embayment—A continental border area that has sagged concurrently with deposits so that an unusually thick section of sediment results.

emplaced—Moved to a particular position. Said of igneous rocks.

eolian—Formed or eroded by the wind.

epicenter—The point on the earth's surface directly above the focus of an earthquake.

erosion—The process by which sediments and rocks are moved and redeposited by water, wind, gravity, or ice.

escarpment—A steep slope at the edge of high ground.

estuary—The partly enclosed tidal mouth of a river where fresh water comes in contact with seawater.

eustatic—Of or pertaining to the worldwide sea level.

exotic terrain—Land that is introduced into another region. Not native to the place where it is found.

Fall Zone—An area characterized by rapids where streams make a sudden descent, as from the Piedmont to the Coastal Plain.

fault—A fracture of the earth's crust along which the opposite sides have been displaced relative to one another.

felsic rock—A light-colored rock that is rich in quartz and feldspar, and poor in iron and magnesium.

floodplain—That portion of a river valley adjacent to the river channel that is built of sediments and covered with water when the river overflows its banks at flood stages.

floodtide delta—Shoals of sediments formed in the backbarrier lagoon as incoming tides pass through an inlet.

fluvial—Of or pertaining to rivers. Produced by river action.

focus—The point on a fault at which movement occurs, causing an earthquake.

foliation—The banded or laminated structure of metamorphic rocks.

foraminifera—A group of mostly marine, unicellular organisms, most of which have shells made of calcite.

foreshore—The zone of the shore or beach that is regularly covered and uncovered by the rise and fall of the tide.

formation—A body of rock strata that consists predominantly of a certain rock type or a combination of rock types.

fossil—Remains or traces of once-living organisms preserved in the earth's rocks or sediments.

fullers earth—A highly absorbent clay.

genus (plural genera)—The major biological subdivision of a family or subfamily in a class of plants and/or animals.

geography—The study of all aspects of the earth's surface including natural and political divisions of countries, the distribution and differentiation of areas, and the relationship of humans to their environment. The topographical features of a region.

geology—The study of the rocks of the earth and the changes that occur over time.

geomorphic—Of or pertaining to the form of the earth or its surface features.

geomorphology—The branch of geology that studies surface landforms and the changes that take place in their evolution.

geosyncline—A major downwarp in the earth's crust, usually more than 620 miles in length, in which sediments accumulate.

glyptodont—A Pleistocene toothless mammal related to the armadillo.

graben—A trenchlike structure bounded by parallel normal faults. A rift.

granite—A crystalline, intrusive igneous rock made up of minerals that include quartz, mica, and feldspar.

gravity anomaly—Difference between theoretical, calculated, and observed terrestrial gravity.

greenhouse effect—A natural effect that traps heat in the atmosphere near the earth's surface, heightened by high levels of carbon dioxide, water vapor and nitrous oxide, methane and chlorofluorocarbons, and ozone.

greenschist—The characteristic of rocks that results from low-grade regional metamorphism.

graywacke—A coarse sandstone, usually dark, containing feldspar and fragments of rock, various minerals, and interstitial clay.

halophytes—Plants that grow in salty or alkaline soil.

hammock—An elevated, well-drained tract of land rising above the general level of a marshy region.

hardpan—Hard, impervious, and often clayey layer of soil at or just below the surface, produced by cementation of soil particles by relatively insoluble materials such as silica, iron oxide, or organic matter.

headwaters—The upper tributaries of a river.

highlands—A plateau or mountainous elevated region of a continent.

humus—A type of soil rich in organic matter.

hydrocarbon—Organic compounds such as oil and gas that contain only hydrogen and carbon.

Iapetus Ocean—The proto-Atlantic Ocean that opened in the Cambrian and closed at the end of the Permian as Pangaea formed.

igneous rocks—Rocks formed by the solidification of a hot fluid called magma.

impermeable—Having a texture that does not allow water to pass.

index fossil—A characteristic fossil that identifies and dates the strata in which it is found.

inlet—A small, narrow opening in a shoreline through which water penetrates into the land. A narrow waterway through a barrier island leading to a bay or lagoon.

inshore—A zone of variable width extending from the low-water shoreline through the breaker zone. The shore face.

interstream divide—The high areas between streams.

island arc—A linear or arcuate-shaped chain of volcanic islands formed at a convergent plate boundary.

isostatic rebound—The property of flotational adjustment made among segments of the lithosphere. For example, a depressed lighter continental crust rebounds vertically over time to float atop denser oceanic crust.

isotope—Any of two or more forms of a chemical element having the same number of protons, but having a different number of neutrons.

kaolin—A soft, white clay composed principally of kaolinite, much used in making ceramics and as a coating on paper.

karst—A type topography formed over limestone, dolomite, or gypsum characterized by sinkholes, caves, and underground drainage.

lagoon—A sound, channel, or bay partly or completely separated from the sea by a reef or barrier island.

lamprey—A jawless fish of the class Agnatha.

lapilli—Volcanic cinders less than an inch in diameter.

Laurentia—The early continent that was to become North America. Proto-North America.

liquefaction—The process of turning soil briefly to a liquid due to seismic activity in water-filled soil.

lithology—The systematic description of rocks relative to their mineral composition and texture.

littoral drift—The movement of water and sand along a shore as currents move landward at an angle. The longshore current.

lode—A mineral deposit located in solid rock.

longshore current—A current within the surf zone that flows parallel to the coast.

mafic rock—Rock composed mainly of dark iron and magnesium-rich minerals.

magma—Molten rock that originates in the earth's upper mantle and forms igneous rocks when cooled.

magnetic anomaly—Any departure from the earth's magnetic field as a whole.

mainland—The shore of a region, as distinguished from adjacent islands.

mantle—The largest portion of the earth's layers located between the core and the crust.

marl—A friable, earthy sedimentary deposit consisting of clay and calcium carbonate.

marsh—An area of low, wet land, often treeless and periodically covered with water.

marsupial—Pouched nonplacental mammals such as opossums, kangaroos, and wombats.

mélange—A chaotic mixture of broken and jumbled rock above a subduction zone.

melt—A body of rock liquified by heat.

member—A body of rock having a high degree of lithologic sameness.

mercury amalgamation process—In gold refining, a complex chemical process that uses mercury to isolate gold particles.

metamorphic—Characterized by the changes in mineralogy and texture imposed on a rock by the pressure, temperature, and chemical environment below the earth's surface.

metamorphic gradient—The intensity of metamorphism, measured by the degree of difference between the parent rock and the metamorphic rock.

metasediment—A previously sedimentary rock that has been altered only slightly by low pressure and heat.

metavolcanics—volcanic rock that has been lightly metamorphosed.

mid-oceanic ridge—The site where magma rises to the surface as two plates move apart at a divergent plate boundary.

mineralization—The process of converting or being converted into a mineral when magma cools. The process of replacing the organic parts of a body by minerals such as wood into stone.

Modified Mercalli Scale—An earthquake intensity scale having twelve divisions ranging from I (not felt by people) to XII (physical damage nearly total).

mollusk—Any invertebrate of the phylum Mollusca including chitons, snails, bivalves, squids, and octopuses. Typically have a calcareous shell that encloses a soft, unsegmented body.

monadnock—A hill or mountain rising conspicuously above the general level of a peneplain. An isolated remnant of rock.

monzonite—A plutonic rock intermediate in composition between syenite and diorite, containing approximately equal parts of alkali feldspar and plagioclase, little or no quartz, and quite often augite as the main mafic mineral.

mosasaur—A cretaceous seagoing lizard that reached 30 feet in length.

mylonitic zone—The area near a fault where the rocks at the point of contact have been milled into a fine-grained breccia.

nannofossil—A marine (usually algal) fossil smaller than a microfossil.

oceanic crust—The dense, thin crust underlying the ocean basins composed of rocks containing silica and magnesium.

ore—A rock or mineral deposit from which a valuable metallic element occurs in high enough concentration to make mining economically feasible.

organic—Characteristic of living organisms.

orogeny—The processes involved in mountain building.

outcrop—That part of a geological formation or structure that appears at the earth's surface.

overburden—Materials of any type that overlie a deposit of useful materials, such as ore or coal, that must be removed.

overthrust—A low-angle thrust fault that covers many miles.

overwash fan—The wave-carried sediment washed into the inlet or marsh side of a barrier island, especially during storms.

oxbow lake—A bow-shaped lake formed by cutoff meanders in an abandoned river channel.

paleontology—The study of fossils.

paleoplacer—A sedimentary deposit that contains valuable minerals, such as gold, that is found in a former riverbed or streambed.

paleoseismology—The study of ancient earthquakes.

Pangaea—The supercontinent formed from all the continents that came together in the Late Paleozoic.

passive margin—A continental margin that has a thick, relatively undeformed sedimentary boundary with only limited tectonism.

peat—Highly organic soil composed of partially decomposed vegetable matter found in marshy or damp regions.

pelecypod—Any aquatic mollusk belonging to the class Pelecypoda, such as clams, characterized by a bilaterally symmetrical bivalve shell, a hatchet-shaped foot, and sheetlike gills.

peneplain—A low, eroded, nearly featureless plain uplifted to form a plateau with renewed exposure to erosion.

permeable—Having a texture that allows water to pass through.

phosphates—Sedimentary rock deposits containing phosphorus, which are mined for fertilizer and other chemical products.

photosynthesis—The process by which plants use water, carbon dioxide, inorganic salts, and sunlight to produce complex organic materials.

phyllite—A slaty metamorphic rock having lustrous cleavage planes due to small scales of mica.

phylum (plural phyla)—The major primary subdivision of the animal kingdom.

physiography—The study of the genesis and evolution of landforms.

pine barrens—Level or slightly rolling land usually with sandy, relatively infertile soil and few trees.

placental mammal—A member of the largest group of mammals that have placentas and nourish their young to term internally for a relatively long period of time.

placer gold—Gold found concentrated at the surface, often in streams, which has been deposited from weathered debris.

plain—An extensive region of level or gently undulating land.

plate—A rigid, thin segment of the earth's crust, oceanic or continental, which differs in chemical composition and tectonic forms.

plate tectonics—A theory that explains movements of continents and changes in the earth's crust caused by internal forces.

playa lake—A shallow lake in an arid region that evaporates during droughts and reforms when water becomes available.

plesiosaur—A large Jurassic to Cretaceous marine reptile having a small head, long neck, four paddlelike limbs, and short tail.

pluton—A large, intrusive igneous rock mass formed at depth in the crust.

pocosin—Freshwater wetland ecosystem characterized by broad-leaved evergreen shrubs or low trees, commonly including pond pine, and often growing on highly organic soils that have developed in areas of poor drainage.

protolith—The material from which a rock is made. For example, sandstone is the protolith of quartzite, shale is the protolith of slate.

protozoans—Single-celled organisms of the phylum Protista, including organisms such as amoeba or paramecium. Among the earliest of lifeforms.

pyroclastic—Pertaining to broken rock material formed by volcanic explosion or aerial expulsion from a volcanic vent.

radioactive—Material that undergoes the emission of some of its energetic particles, and so changes into another element over time.

radiocarbon dating—A method of finding the age of organic substances by determining the amount of carbon-14 present.

radiometric dating—A method that uses the decay of radioactive elements, such as uranium, to determine the absolute age of rocks and minerals.

red beds—Sedimentary strata composed mainly of sandstone, shale, and siltstone colored red by iron oxide.

refraction—The bending or deflection of a wave as its speed is changed while passing through a medium.

regional metamorphism—Metamorphism that occurs over a wide area and is caused by the deep burial or strong tectonic forces of the earth, such as occurs in continental collisions.

replacement deposit—A replacement of ore minerals by mineralized hot water solutions that have first dissolved the original mineral.

residuum—Material that is left over. Residue.

Richter Scale—The range of numerical values of earthquake magnitude on a scale of 1 to 10, used to compare the amounts of energy released by an earthquake. *See also* Modified Mercalli Scale.

ridge—A steep, relatively narrow elevation.

rift—A valley along the trace of a fault within a divergent boundary. The central cleft in the crest of a mid-oceanic ridge.

rift basin—A series of grabens formed at a divergent plate boundary.

rift zone—A system of crustal fractures and faults at a divergent plate boundary.

runnels—Channels where water runs between successive beach ridges.

rupture—A fracture, as when rock breaks along a fault during an earthquake.

salinity—The total quantity of dissolved salts in seawater.

sandbanks—Bars or ridges of sand built up to or near the surface of water by currents in a river or by wave action in coastal waters. Tidal shoals.

sandbar—A ridge of sand built up to or near the surface of water by currents or waves.

sandblows—Craterlike areas formed when soil under pressure during an earthquake quickly rises to the surface.

saprolite—A soft, earthy, clay-rich, and thoroughly decomposed rock formed in place as a result of the chemical weathering of igneous or metamorphic rocks.

savannah—level land, often having wet soil part of the year that supports low vegetation such as grasses and scattered bushes and trees.

scarp—*See* escarpment.

schist—A medium- to coarse-grained, foliated metamorphic rock.

sea island—A coastal island having a core that is made up of mainland materials inundated as sea levels rose and/or the mainland sank.

Sea Island Complex—The sea islands present along the Atlantic coast from Charleston to north Florida.

sediment—Solid material, both organic and inorganic, formed by weathering and transported by erosional forces.

sedimentary rocks—Rocks formed from the compaction and/or cementation of mineral grains carried by physical agents such as wind, water, ice, or gravity, or by chemical or biological agents.

seismicity—The frequency, intensity, and distribution of earthquakes in a given area.

seismic reflection—A mode of seismic mapping in which artificial seismic waves are generated into the earth and recorded by instruments forming images of the substrate as they are reflected and refracted by underlying rock.

seismic wave—A body and surface wave produced by earthquakes, or artificially by explosion, which travels both through the earth and along the surface.

seismic zone—An area where earthquakes occur.

seismometer—An instrument for magnifying and monitoring the motions of the earth's surface that are caused by seismic waves.

shear zone—A fracture zone caused by stresses that move one part of a rock body past the adjacent part, generally resulting in pulverized material along its surface.

shoals—A relatively shallow place in a body of water. A submerged or partly submerged ridge, bank, or bar of sand in a body of water.

shocked quartz—Silica that shows microscopic changes, such as crustal realignment or fracturing, from the effects of shock waves produced by an explosion or meteorite impact.

silicates—A large group of minerals, including quartz and feldspar, composed of silicon, oxygen, and one or more other elements.

sill—A thin, horizontal igneous intrusion emplaced parallel to overlying rocks.

slippage—Relative movement of formerly adjacent rock layers, as on opposite sides of a fault.

sluice—An artificial channel built to direct water over sediments in order to trap gold particles.

spit—A long, narrow ridge of sand deposited by a current as the current loses velocity.

stratigraphy—The science of the description, correlation, and classification of strata in sedimentary rocks, including the interpretation of those rock layers.

stratum (plural strata)—A single sedimentary bed or layer.

stromatolites—Precambrian to Recent microfossils similar to blue-green algae that secrete a calcareous skeleton that forms characteristic mushroom-shaped limestones at seacoasts.

subduction—The descent of one crustal plate under another.

subsoil—The stratum of earth immediately under the subsurface or topsoil.

swamp—Low, spongy land generally saturated with moisture which often has a growth of trees and other vegetation. Generally unfit for agriculture.

swash—The rush of water onto a beach after the breaking of a wave.

syncline—A *U*-shaped fold of rock that usually concaves upward.

synform—A large synclinal structure where the stratigraphic sequence is unknown.

tailings—Those portions of ore that are regarded as too poor to refine. Residue ore.

tannins—Any of a group of astringent vegetable compounds that leach out of leaves and bark to discolor river water.

tapir—A hoofed, four-legged mammal that somewhat resembles swine and has a long, flexible snout.

tectonics—The forces or conditions within the earth that cause movements of the crust such as earthquakes, folds, and faults.

terrain—The area or surface over which a particular group of rocks is prevalent.

terrestrial—Of or pertaining to the earth.

terrigenous—Coming from the land, as in sediments that are eroded from land to the ocean.

Therapsid—The reptile group from which the mammal orders descended.

thrust sheet—The block of rock that is pushed at a low angle by a thrust fault.

tidal range—The difference between the level of water at high tide and low tide.

topography—The physical features of a region. The shape of the earth's surface.

trench—A long, narrow, deep trough in the sea floor caused by the subduction of oceanic crust under a continental crust or another oceanic crust.

trilobites—An extinct group of marine arthropods that lived throughout the Paleozoic.

troy ounce—A unit of weight used to measure precious metals and gems. Twelve troy ounces equals 1 pound.

tuff—Rock produced by the deposition of pyroclastic volcanic fragments usually smaller than .156 inches (4 millimeters).

turbidites—A bottom-flowing current laden with sediments that flows downslope under water.

Turritella—A genus of Triassic to Recent slender, small, whorled mollusk.

ultramafic rock—A dark igneous rock (such as dunite, pyroxenite, peridotite, or amphibolite) that consists mainly of mafic minerals, but contains less than 10 percent feldspar.

unconformity—A surface between two strata that represents an interval of time in which deposition stopped, erosion removed some sediment and rock, and deposition then resumed.

vascular—Having vessels or ducts that convey fluids such as sap or blood.

velociraptor—A carnivorous ornithischian dinosaur with rapierlike claws on its hind limbs.

water eddy—*See* eddy.

watershed—The area drained by a river or stream.

water table—The upper surface of a permeable body of rock between a zone of saturation and a zone of aeration. Ground water level.

wetlands—Area of land, excluding streams, lakes, or open oceans, that remains wet for a large portion of the year with salt or fresh water.

zeolite—A class of silicates that contains water within the crystal structure, formed by alteration of other silicates at low temperature and pressure.

RESOURCES

FEDERAL AGENCIES

United States Geological Survey
P.O. Box 25046
Denver, Colorado 80225
303-236-5900
The research office of the Survey.

United States Geological Survey
12201 Sunrise Valley Drive
Reston, Virginia 22092
703-648-4000
Books, pamphlets, photographs, teaching aids, and maps can be purchased.

United States Department of Agriculture, Forest Service
4931 Broad River Road
Columbia, South Carolina 29210
803-561-4091
Maps and brochures of the national forest lands are available.

Natural Resources Conservation Service, United States Department of Agriculture
1835 Assembly Street, Room 950

Columbia, South Carolina 29201
803-253-3975
Call the district offices in each county of South Carolina.

STATE AGENCIES

South Carolina Department of Natural Resources
1000 Assembly Street
P.O. Box 167
Columbia, South Carolina 29202
803-734-3888
For information on the Carolina bays and other protected sites, contact the Non-Game and Heritage Trust, 803-734-3893.

South Carolina Geological Survey
5 Geology Road
Columbia, South Carolina 29210
803-896-7713
Books, pamphlets, rock and mineral samples, and maps can be purchased.

Land Resources and Conservation Division
South Carolina Department of Natural Resources
Cartographic Information and Map Sales
2221 Devine Street, Suite 222
Columbia, South Carolina 29205
803-734-9100
Maps can be purchased; geology brochures and map interpretation advice available.

ACE Basin Preserve
South Carolina Department of Natural Resources
Route 1, Box 25
Green Pond, South Carolina 29446
803-762-5400; 803-889-3084

South Carolina Department of Parks, Recreation, and Tourism
1205 Pendleton Street
Columbia, South Carolina 29201
803-734-0122
Guide to state parks, South Carolina state highway map, South Carolina travel guides, and South Carolina waterfall guide: "Finding the Falls" are available.

South Carolina Institute of Archeology and Anthropology
1321 Pendleton Street
Columbia, South Carolina 29208
803-734-0566
A hobby diving license needs to be obtained to collect fossils in South Carolina public waters, including rivers, lakes, and up to 3 miles offshore in the ocean. Call or write the Institute for license forms.

South Carolina Sea Grant Consortium
287 Meeting Street
Charleston, South Carolina 29401
803-727-2078
Material on coastal environments. Publications list available.

EDUCATIONAL CENTERS

Savannah River Site Ecology Laboratory
P.O. Drawer E
Aiken, South Carolina 29802
803-725-2472
School field trips can be scheduled.

Center for Science Education
John Carpenter, Director
University of South Carolina
Columbia, South Carolina 29208
803-777-6920

The Center provides help to educators who teach science in
South Carolina. The Center distributes videos, books, rock, and
mineral samples, and offers workshops, advice, and courses
throughout the year to educators. The Center has produced field
trip guides for teachers including such sites as the Congaree
Swamp National Monument, Hunting Island State Park,
Peachtree Rock, and Forty Acre Rock.

Thomas Cooper Map Library
University of South Carolina
Greene Street
Columbia, South Carolina 29208
803-777-2802
One of the most complete map libraries in the Southeast.
Noncirculating collection available for research. Open
Monday–Friday: 8:00 A.M. to 5:00 P.M.

Coastal Zone Education Center
University of South Carolina at Beaufort
285 Sawmill Creek Road
Bluffton, South Carolina 29910
803-837-4848
Provides field trips for schools and residential programs on
barrier islands and coastal studies for the public.

**Seismic Monitoring Network and Earthquake
Education Center**
Dr. Joyce Bagwell, Director
Charleston Southern University
9200 University Boulevard
P.O. Box 118087
Charleston, South Carolina 29423-8087
803-863-8088
Publications available: "How to Survive an Earthquake,"
guidebook for developing a school earthquake program, and
brochure about earthquakes. The Center also loans films, slides,
books, and 3-D earth science models for demonstrations. These

are available free to teachers, students, and interested citizen groups.

Bellefield Nature Center
U.S. Highway 17
P.O. Box 1630
Georgetown, South Carolina 29442
803-546-3623
Located north of Georgetown. Information and tours of estuarine and forest environments.

Tom Yawkey Wildlife Center
South Carolina Department of Natural Resources
Route 2, Box 181
Georgetown, South Carolina 29440
803-546-6814
Free tours of the refuge (South Island) only on one weekday (usually Tuesday or Wednesday): 3:00–6:00 P.M. Call for appointment.

Roper Mountain Science Center
501 Roper Mountain Road
Greenville, South Carolina 29615
803-281-1188
Off U.S. Interstate 385. Open to public second Saturday of the month; nature trail daily; school field trips can be arranged.

Tugaloo Environmental Education Center
351 Teec Drive
Westminster, South Carolina 29693
803-647-4930; 1-800-741-1139 to make reservations
Classes for student groups and teacher education.

Spartanburg Science Center
Spartanburg Arts Center
385 South Spring Street

Spartanburg, South Carolina 29306
803-583-2777
Saturday programs for the public; school programs and science camps.

MUSEUMS

South Carolina State Museum
301 Gervais Street
Columbia, South Carolina 29202
803-737-4921
Rocks, fossils, and displays; demonstrations and discovery labs for schools.

Charleston Museum
360 Meeting Street
Charleston, South Carolina 29403
803-722-2996
Rocks, minerals, and fossils.

Laurence L. Smith Geology Museum
McKissick Museum
University of South Carolina
Columbia, South Carolina 29208
803-777-7251

Clemson University
Geology Department
221 Brackett Hall
Clemson, South Carolina 29634
803-656-4481
Rocks, fossils, minerals, and large gemstone display.

Furman University
Science Building, Plyler Hall
Poinsett Highway (U.S. Highway 276)

Greenville, South Carolina 29613
803-294-2052
Fossils and rocks.

Charleston Southern University
Ashby Hall
9200 University Blvd.
P.O. Box 118087
Charleston, South Carolina 29423
803-863-7000
Minerals, rocks, and fossils.

The Museum
106 Main Street
Greenwood, South Carolina 29648
803-229-7093
Minerals, rocks, and fossils.

Museum of York County
4621 Mount Gallant Road
Rock Hill, South Carolina 29730
803-329-2121
Nature trail and special tours for groups.

Discovery Place
301 North Tryon Street
Charlotte, North Carolina 28202
704-845-3882
Minerals, rocks, and fossil displays.

MINING FIELD TRIPS

Gold Hill Mine
Interstate Highway I-77 (Carowinds Exit)
2707 U.S. Highway 21 (business)
Fort Mill, South Carolina 29715

803-548-6463
Panning for gems and gold.

Cottonpatch Gold Mine
41697 Gurley Road
New London, North Carolina 28127
704-463-5797
Wash buckets of gold; equipment provided.

Franklin, North Carolina
U.S. Highway 441
Many gem mines. Call Franklin Area Chamber of Commerce for listings: 1-800-336-7829.

Reed Gold Mine
9621 Reed Mine Road
Stanfield, North Carolina 28163
704-786-8337
20 miles north of Charlotte off N.C. Highway 24, 2 miles east of U.S. Highway 601. Museum display, mine tour, film, and gold panning.

ROCK SHOPS

Shaw's Rockart and Crafts
2218 Holland Street
West Columbia, South Carolina 29169
803-794-5794

Beckham's Barn
1751 Kennerly Road
Irmo, South Carolina 29063
803-781-3589

Dixie Gem
2517 Poinsett Highway

Greenville, South Carolina 29609
803-235-7309
2 miles south of Furman University on U.S. Highway 276.

Lowcountry Geologic
518 Woodland Shores Road
Charleston, South Carolina 29412
803-795-2956

SCIENCE SUPPLIES

Wards Natural Science Establishment
5100 West Henrietta Road
P.O. Box 92912
Rochester, New York 14692-9012
1-800-962-2660

Carolina Biological Supply
2700 York Road
Burlington, North Carolina 27215
910-584-0381

EARTH SCIENCE CLUBS

Columbia Gem and Mineral Society
P.O. Box 6333
Columbia, South Carolina 29620
803-256-3713

South Carolina Earth Antiquities Society
P.O. Box 42234
Columbia, South Carolina 29206
803-782-0011

Low Country Gem and Mineral Society
1857 Sandcroft Drive

Charleston, South Carolina 29407
803-571-7172

Charles Town Mineral and Lapidary Club
P.O. Box 902
Johns Island, South Carolina 29455
803-559-3459

Pendleton District Gem and Mineral Society
258 East Bear Swamp Road
Walhalla, South Carolina 29691
803-638-5114

Charlotte Gem and Mineral Club
P.O. Box 10233
Charlotte, North Carolina 28212
704-552-0389

Western South Carolina Gem and Mineral Society
37 North Avondale Drive
Greenville, South Carolina 29615
803-233-0575

Aiken Gem and Mineral Society
7 Knollwood Boulevard
North Augusta, South Carolina 29841
803-278-5878

Myrtle Beach Fossil Club
2711 Lee's Landing Circle
Conway, South Carolina 29526
803-347-7592

Grand Strand Fossil Club
519 7th Avenue South

Surfside Beach, South Carolina 29575
803-238-1083

ASSOCIATIONS

Carolina Geological Society
Duke University
Department of Geology
206 Old Chemical Building
Durham, North Carolina 27706
Yearly field trips.

South Carolina Nature Conservancy
2231 Devine Street, Suite 100
P.O. Box 5475
Columbia, South Carolina 29250
803-254-9049

Sierra Club
South Carolina Chapter
1314 Lincoln Street
P.O. Box 2388
Columbia, South Carolina 29202
803-256-8487

Mining Association of South Carolina
P.O. Drawer 1368
Irmo, South Carolina 29063
803-772-5354
Maps and brochures on state mining industries.

South Carolina Coastal Council
4130 Faber Place, Suite 300
Charleston, South Carolina 29405
803-744-5838

Coastal Conservation League
P.O. Box 1765
Charleston, South Carolina 29402
803-723-8035

MAGAZINES

South Carolina Wildlife
P.O. Box 167
Columbia, South Carolina 29202

Earth
Kalmbach Publishing Co.
21027 Crossroads Circle
P.O. Box 1612
Waukesha, Wisconsin 53187

Natural History Magazine
American Museum of Natural History
Central Park West at 79th Street
New York, New York 10024

National Geographic
National Geographic Society
P.O. Box 2895
Washington, D.C. 20077-9960

Science News
P.O. Box 1925
Marion, Ohio 43305

Geotimes
American Geological Institute
4220 King Street
Alexandria, Virginia 22302-1507

RECOMMENDED BOOKS

The Geology of the Carolinas, J. Wright Horton, Jr., and Victor A. Zullo, eds. (Knoxville: University of Tennessee Press, 1991)

The Making of a Continent, Ron Redfern (Times Books, New York, 1983)

The Evolution of North America, Philip B. King (Princeton University Press, 1977, rev.)

Written in Stone, Chet and Maureen E. Raymo (Globe Pequot Press, Chester, Conn., 1989)

Earth, Frank Press and Raymond Siever (W. H. Freeman and Company, New York, 1986)

Wonderful Life, Stephen Jay Gould (W. W. Norton and Company, New York, 1989)

The Geologic Evolution of North America, Colin Stearn (John Wiley and Sons, New York, 1979)

Hiking South Carolina Trails, 3rd ed., Allen de Hart (Globe Pequot Press, Chester, Conn., 1994)

Plate Tectonics for Introductory Geology, John R. Carpenter and Philip Astwood (Kendall/Hunt Publishing, Dubuque, Iowa, 1983)

BIBLIOGRAPHY

American Geological Institute. *Dictionary of Geologic Terms.* Garden City, N.J.: Anchor Press, 1976.

Anderson, Alan H. *The Drifting Continents.* New York: G. P. Putnam and Sons, 1971.

Arduini, Paolo, and Giorgio Teruzzi. *Guide to Fossils.* New York: Simon and Schuster, 1986.

Assembly of Mathematical and Physical Sciences, (U.S.) Geophysics Study Committee. *Continental Tectonics.* Washington, D.C., National Academy of Sciences, 1980.

Astwood, Philip. University of South Carolina Center for Science Education. Personal communication, 1993.

Bagwell, Dr. Joyce. Charleston Southern University. Personal communication, 1992.

Bates, Robert L., and Julia Jackson, eds. *Dictionary of Geological Terms.* New York: American Geological Institute, Doubleday, 1984.

Bell, Henry, III, et al. *Geology of the Piedmont and Coastal Plain near Pageland, South Carolina and Wadesboro, North Carolina.* Columbia: South Carolina Geological Survey, 1974.

Bennett, Stephen. "Landmarks of Mystery." *South Carolina Wildlife,* (September/October 1989).

————, and John B. Nelson. *Distribution and Status of Carolina Bays in South Carolina.* Columbia: South Carolina Non-Game and Heritage Trust Publications, South Carolina Wildlife and Marine Resources Department, 1991.

Blagden, Tom, Jr. *South Carolina's Wetland Wilderness: The ACE Basin.* Englewood, Colo.: Westcliffe Publishers, 1992.

Bliley, Daniel, and David Burney. "Late Pleistocene Climatic Factors in the Genesis of a Carolina Bay." *Southeastern Geology* (December 1988).

Bollinger, G. A. "Studies Related to the Nature of Seismicity at Charleston, South Carolina." *Studies Related to the Charleston, South Carolina, Earthquake of 1886—Tectonics and Seismicity.* United States Geological Survey Professional Paper 1313. Reston, Va.: USGS, 1983.

Bond, Gerard C., et al. "Breakup of a Supercontinent between 625 Ma and 555 Ma: New Evidence and Implications for Continental Histories." *Earth and Planetary Science Letters* 70 (1984): 325–45.

Burchfiel, B. Clark. "The Continental Crust." *Scientific American* (September 1983).

Butler, J. Robert. *Geology and Mineral Resources of York County, South Carolina.* Bulletin 33. Columbia: South Carolina State Development Board, 1966.

————. "Metamorphism." In *The Geology of the Carolinas.* J. Wright Horton, Jr., and Victor A. Zullo, eds. Knoxville: University of Tennessee Press, 1991.

————, and Donald T. Secor, Jr. "The Central Piedmont." In *The Geology of the Carolinas.* J. Wright Horton, Jr., and Victor A. Zullo, eds. Knoxville: University of Tennessee Press, 1991.

Campbell, Bob. "Pearl of the Lowcountry." *South Carolina Wildlife* (September/October 1989).

Campbell, William P. *Appalachian Gold.* Johnson City, Tenn.: Overmountain Press, 1976.

Carpenter, John R., and Philip M. Astwood. *Plate Tectonics for Introductory Geology.* New York: Kendall/Hunt Publishing Co., 1983.

Case, Gerard. *A Pictorial Guide to Fossils.* New York: Van Nostrand Reinhold Co., 1982.

Cherrywell, Christopher, and Keith Tockman. "Gold Production in South Carolina." *South Carolina Geology* 34 (1992).

Chesterman, Charles W. *Audubon Society Field Guide to North American Rocks and Minerals.* New York: Alfred A. Knopf, 1978.

Chowns, T. M., and C. T. Williams. "Pre-Cretaceous Rocks Beneath the Georgia Coastal Plain—Regional Implications." *Studies Related to the Charleston, South Carolina, Earthquake of 1886—Tectonics and Seismicity.* United States Geological Survey Professional Paper 1313. Reston, Va.: USGS, 1983.

City of Charleston. *The Yearbook of 1886.* Charleston, S.C.: 1886.

Colquhoun, Donald J. *Terrace Sediment Complexes in Central South Carolina.* Columbia: University of South Carolina Press, 1965.

————. "Geomorphology of River Valleys in the Southeastern Atlantic Coastal Plain." *Southeastern Geology* 7 (1966).

————. *Geomorphology of the Lower Coastal Plain of South Carolina.* Columbia: South Carolina State Development Board, 1969.

Cook, Frederick A. "Thin-skinned Tectonics in the Crystalline Appalachians: COCORP Seismic-reflection Profiling of the Blue Ridge and Piedmont." *Geology* (December 1979).

————. "The Southern Appalachians and the Growth of Continents." *Scientific American* (October 1980).

Cooke, C. Wythe. *Carolina Bays and the Shapes of Eddies.* United States Geological Survey Professional Paper 254-I. Washington, D.C.: USGS, 1954.

Daniels, David L., Isidore Zietz, and Peter Popenoe. "Distribution of Subsurface Lower Mesozoic Rocks in the Southeastern United States, as Interpreted from Regional Aeromagnetic and Gravity Maps." *Studies Related to the Charleston, South Carolina, Earthquake of 1886—Tectonics and Seismicity.* United States Geological Survey Professional Paper 1313. Washington, D.C.: U.S. G.P.O., 1983.

Davenport, Jim. "Gold Diggers, Is There Glitter in South Carolina's Mines?" *The State.* March 19, 1990.

Dennis, Allen Johnson. "Tectogenesis of an Accreted Terrain: The Carolina Arc in the Paleozoic. " Ph.D. diss., University of South Carolina, 1989.

Dickey, Beth. "Thar's Still Gold in Them Thar Hills." *Sandlapper* (January 1990).

Dietz, Robert, and John C. Holden. "The Breakup of Pangaea." *Scientific American* (October 1970).

————. "Geosynclines, Mountains, and Continent-building." *Scientific American* (March 1972).

Dillon, William P., Kim D. Klitford, and Charles K. Paull. "Mesozoic Development and Structure of the Continental Margin off South Carolina." *Studies Related to the Charleston, South Carolina Earthquake of 1886—Tectonics and Seismicity.* United States Geological Survey Professional Paper 1313. Reston, Va.: USGS, 1983.

Dillon, William P., and Peter Popenoe. "The Blake Plateau Basin and Carolina Trough." In *The Geology of North America: The Atlantic Margin,* vol. I-2, R. E. Sheridan and J. A. Grow, eds. Denver, Colo.: Geological Society of America, 1988.

DuBar, Jules R. *Neogene Stratigraphy of the Lower Coastal Plain of the Carolinas.* Myrtle Beach, S.C.: Atlantic Coastal Plain Geological Association, 1971.

Eckert, Allan W. *Earth Treasures: The Southeastern Quadrant.* New York: Harper and Row, 1987.

Ellerbe, Clarence M. *South Carolina Soils and their Interpretations for Selected Uses.* Columbia: South Carolina Land Resources Conservation Commission, 1974.

Fairey, Daniel. *South Carolina's Geologic Framework.* Columbia: South Carolina Land Resources Commission, 1977.

———. *South Carolina's Land Resources: A Regional Overview.* Columbia: South Carolina Land Resources Commission, 1992.

Feiss, P. Gregory, et al. "Mineral Resources of the Carolinas." In *The Geology of the Carolinas.* J. Wright Horton, Jr., and Victor A. Zullo, eds. Knoxville: University of Tennessee Press, 1991.

Field, Michael E., and David B. Duane. "Post-Pleistocene History of the United States Inner Continental Shelf: Significance to Origin of Barrier Islands." *Geological Society of America Bulletin* 87 (May 1976).

Fisher, John J. "Barrier Island Formation: Discussion." *Geological Society of America Bulletin* 79 (October 1968).

Fortey, Richard. *Fossils: The Key to the Past.* New York: Van Nostrand Reinhold Co., 1982.

Gaffney, Eugene S. *Dinosaurs.* New York: Golden Press, 1990.

Gillen, Con. *Metamorphic Geology.* London: George Allen and Unwin, 1982.

Gohn, G. S., ed. *Studies Related to the Charleston, South Carolina Earthquake of 1886—Tectonics and Seismicity.* United States Geological Survey Professional Paper 1313. Reston, Va.: USGS, 1983.

Gould, Stephen Jay. *Ever Since Darwin, Reflections in Natural History.* New York: W. W. Norton and Co., 1973.

———. *The Flamingo's Smile.* New York: W. W. Norton and Co., 1985.

———. *Wonderful Life.* New York: W. W. Norton and Co., 1989.

Hamilton, Robert, et al. "Land Multichannel Seismic-Reflection Evidence for Tectonic Features near Charleston, South Carolina." *Studies Related to the Charleston, South Carolina, Earthquake of 1886—Tectonics and Seismicity.* United States Geological Survey Professional Paper 1313. Reston, Va.: USGS, 1983.

Harpers New Monthly Magazine (August 1857).

Harris, Leonard D., and Kenneth C. Bayer. "Sequential Development of the Appalachian Orogeny above a Master Decollement—A Hypothesis." *Geology* (December 1979).

Hatcher, Robert, Jr. "Developmental Model for the Southern Appalachians." *Geological Society of America Bulletin* 83, no. 9 (1972).

————. *Introduction to the Geology of the Eastern Blue Ridge of the Carolinas and Nearby Georgia*. Columbia: Carolina Geological Survey, South Carolina Development Board, 1976.

————, ed. *The Appalachian-Ouachita Orogeny in the United States*. Boulder, Colo.: Geological Society of America, 1989.

————, and Stephen A. Goldberg. "The Blue Ridge Geologic Province." In *The Geology of the Carolinas*. J. Wright Horton, Jr., and Victor A. Zullo, eds. Knoxville: University of Tennessee Press, 1991.

————, and George W. Viele. "The Appalachian/Ouachita Orogeny: U.S. and Mexico." In *Perspectives in Regional Geological Synthesis*. Boulder, Colo.: Geological Society of America, 1982.

Hayes, Miles O., and Tim Kana. *Terrigenous Clastic Depositional Environments*. Technical Report No. 11. Columbia: University of South Carolina, Coastal Resources Division, 1976.

————. Personal communication, 1992.

Hays, Walter, and Paula L. Gori, eds. *A Workshop on the 1886 Charleston, South Carolina, Earthquake and its Implications for Today*. Reston, Va.: USGS, 1983.

Heron, S. Duncan, Jr., and Henry S. Johnson. *Radioactive Mineral Resources of South Carolina*. Columbia: South Carolina State Development Board, 1969.

Horner, John R., and James Gorman. *Digging Dinosaurs*. New York: Workman Publishing, 1988.

Horton, J. Wright. *Geologic Investigations of the Kings Mountain Belt and Adjacent Areas in the Carolinas*. Columbia: South Carolina Geological Survey, 1981.

————, and Zullo, Victor A., eds. *The Geology of the Carolinas*. Knoxville: University of Tennessee Press, 1991.

Howe, Jerry T., and Andrew Howard. *Fossil Locations in South Carolina, State Museum Bulletin 3*. Columbia: South Carolina State Museum, 1990.

Hoyt, John H., and Vernon Henry. "Influence of Island Migration on Barrier Island Sedimentation." *Geological Society of America Bulletin* 78 (January 1967).

Hughes, Bill. "Sand Dollars." *The State*. December 17, 1992.

Ingmanson, Dale E., and William J. Wallace. *Oceanography: An Introduction*. Belmont, Calif.: Wordsworth Publishing Co., 1989.

James, Dr. L. Allan. University of South Carolina Geography Department. Personal communication, 1992.

Johnson, H. S. "Geology in South Carolina." *Miscellaneous Report 3*. Columbia: South Carolina State Development Board, 1964.

Johnston, Arch C., and Lisa R. Kanter. "Earthquakes in Stable Continental Crust." *Scientific American* (March 1990).

Jones, Lewis P. *South Carolina, One of the Fifty States.* Orangeburg, S.C.: Sandlapper Publishing Co., Inc., 1985.

Kaczorowski, Raymond T. "The Carolina Bays and Their Relationship to Modern Oriented Lakes." Ph.D. diss., University of South Carolina, 1977.

Kana, Dr. Timothy. Coastal Science and Engineering, Inc. Personal communication, 1993.

King, Philip B. *The Evolution of North America.* Princeton: Princeton University Press, 1977.

Kite, Lucille E. "Stratigraphy of Peachtree Rock Preserve, Southern Lexington County, South Carolina." *Geology* 29, no. 1 (1985).

Kovacik, Charles, and John Winberry. *South Carolina.* Boulder, Colo.: Westview Press, 1987.

Kurten, Bjorn. *The Age of Mammals.* London: Weidenfield and Nicolson, 1971.

———, and Elaine Anderson. *Pleistocene Mammals of North America.* New York: Columbia University Press, 1980.

Leatherman, Stephen P. *Barrier Island Handbook.* College Park, Md.: University of Maryland Press, 1982.

Livingston, Mike. "Lowcountry Tremblor Gives Folks the Shakes." *The State.* August 22, 1992.

Long, Leland Timothy. "The Carolina Slate Belt—Evidence of a Continental Rift Zone." *Geology* 7 (April 1979).

Marple, Ronald, and Pradeep Talwani. "Evidence of Possible Tectonic Upwarping along the South Carolina Coastal Plain from an Examination of River Morphology and Elevation Data." *Geology* 21, no. 7 (July 1993).

Maher, Harmon D., Jr., et al. "The Eastern Piedmont in South Carolina." In *The Geology of the Carolinas.* J. Wright Horton, Jr., and Victor A. Zullo, eds. Knoxville: University of Tennessee Press, 1991.

McCauley, Camilla K., and J. Robert Butler. *Gold Resources of South Carolina. Bulletin 32.* Columbia: South Carolina Geological Survey, 1966.

McKenzie, John C., and John F. McCauley. *Geology and Kyanite Resources of Little Mountain.* Columbia: South Carolina State Development Board, 1968.

McSween, Harry Y., Jr., et al. "Plutonic Rocks." In *The Geology of the Carolinas.* J. Wright Horton, Jr., and Victor A. Zullo, eds. Knoxville: University of Tennessee Press, 1991.

Moorbath, Stephen. "The Oldest Rocks and the Growth of Continents." *Scientific American* (October 1977).

Nance, R. Damian, et al. "The Supercontinent Cycle." *Scientific American* (July 1988).

Neal, Donald W. *Geology of Eocene Sediments around Santee, South Carolina.* Greenville, N.C.: East Carolina University, 1988.

Neal, William J., et al. *Living with the South Carolina Shore.* Durham: Duke University Press, 1984.

Nystrom, Paul G., et al. "Cretaceous and Tertiary Stratigraphy of the Upper Coastal Plain, South Carolina." In *The Geology of the Carolinas.* J. Wright Horton, Jr., and Victor A. Zullo, eds. Knoxville: University of Tennessee Press, 1991.

Olsen, Stanley J. *Fossil Mammals of Florida.* Tallahassee: Florida Geological Survey, 1959.

Overstreet, William C., and Henry Bell. *The Crystalline Rocks of South Carolina.* Washington, D.C.: USGS, 1965.

Pope, Charles. "SRS Tritium Discovered in 4 Wells." *The State.* October 23, 1991.

———. "Coastal Clashes." *The State.* November 22, 1992.

Popenoe, Peter, and Isidore Zietz. "The Nature of the Geophysical Basement beneath the Coastal Plain of South Carolina and Georgia." *Studies Related to the Charleston, South Carolina, Earthquake of 1886—A Preliminary Report.* United States Geological Survey Professional Paper 1028. Reston, Va.: USGS, 1977.

Press, Frank, and Raymond Sieven. *Earth.* New York: W. H. Freeman and Co., 1986.

Rankin, Douglas W. "Introduction and Discussion." *Studies Related to the Charleston, South Carolina, Earthquake of 1886—A Preliminary Report.* United States Geological Survey Professional Paper 1028. Reston, Va.: USGS, 1977.

Raymo, Chet. *Written in Stone.* Chester, Conn.: Globe Pequot Press, 1989.

Redfern, Ron. *The Making of a Continent.* New York: Times Books, 1983.

Rhodes, Frank H. T. *Fossils.* New York: Golden Press, 1962.

Roberts, Bruce. *The Carolina Gold Rush.* Charlotte: McNalley and Loftin, 1972.

Roberts, Nancy. *The Gold Seekers.* Columbia: University of South Carolina Press, 1989.

Rodgers, John. *The Tectonics of the Appalachians.* New York: Wiley Interscience, 1970.

Salda, Luis, et al. "Did the Taconic Appalachians Continue into Southern South America?" *Geology* 20 (December 1992).

Samson, Sara L. "Middle Cambrian Fauna of the Carolina Slate Belt,
    Central South Carolina." Master's thesis, University of South Carolina,
    1984.
Savage, Henry. *The Mysterious Carolina Bays.* Columbia: University of
    South Carolina Press, 1982.
Schwimmer, David R., and Robert H. Best. "First Dinosaur Fossils from
    Georgia, with Notes on Additional Cretaceous Vertebrates from the
    State." *Georgia Journal of Science* 47 (1989).
Sharitz, R. R., and J. W. Gibbons. *The Ecology of Southeastern Scrub
    Bogs (Pocosins) and Carolina Bays: A Community Profile.* Washing-
    ton, D.C.: Fish and Wildlife Service, Division of Biological Services,
    1982.
Sloan, Earle. *Catalogue of the Mineral Localities of South Carolina.*
    Series 4, Bulletin 2. Columbia: South Carolina Geological Survey,
    1908.
Smith, George E., III. "Lithostratigraphic Relationships of Coastal Plain
    Units in Lexington County and Adjacent Areas, South Carolina."
    Master's thesis, University of South Carolina, 1979.
Snoke, Arthur, ed. *Geologic Investigations of the Eastern Piedmont,
    Eastern Appalachians.* Columbia: Carolina Geological Society, South
    Carolina State Development Board, 1978.
Soller, David R., and Hugh H. Mills. "Surficial Geology and Geomor-
    phology." In *The Geology of the Carolinas.* J. Wright Horton, Jr., and
    Victor A. Zullo, eds. Knoxville: University of Tennessee Press, 1991.
South Carolina Geological Survey. *Catalog of South Carolina Mineral
    Producers.* Circular 2. Columbia: South Carolina Geological Survey,
    1988.
————. *Catalog of Geologic Publications, 1989–1990.* Columbia: 1990.
South Carolina Seismic Safety Consortium. *Earthquake Hazards, Risk,
    and Mitigation in South Carolina and the Southeastern United States.*
    Charleston: Citadel, 1986.
Stearn, Colin, et al. *Geological Evolution of North America.* New York:
    John Wiley and Sons, 1979.
Stockton, Robert. *The Great Shock.* Easley, S.C.: Southern Historical
    Press, 1986.
Stokes, William L. *Essentials of Earth History.* Englewood Cliffs, N.J.:
    Prentice-Hall, 1973.
Stratton, Karen, and Alan Weekes. *Field Trip Guide to Peachtree Rock.*
    Columbia: University of South Carolina Center for Science Education,
    1993.
Sullivan, Walter. *Landprints.* New York: Times Books, 1984.

Sutcliffe, Anthony. *On the Track of Ice Age Mammals.* Cambridge: Harvard University Press, 1985.

Talwani, Pradeep. "Current Thoughts on the Cause of the Charleston, South Carolina, Earthquakes." *South Carolina Geology* 29, no. 2 (1985).

———. "Seismotectonics of the Charleston Earthquake" In *Proceedings of the Third U.S. National Conference on Earthquake Engineering.* El Cerrito, Calif.: Earthquake Engineering Research Institute, 1986.

Tewhey, John D. *Geology of the Irmo Quadrangle, Richland and Lexington Counties, South Carolina.* Columbia: South Carolina State Development Board, 1977.

Tuzo, William J., ed. *Continents Adrift, Continents Aground, Readings from Scientific American.* San Francisco: W. H. Freeman and Co., 1976.

VandenDolder, Evelyn M. "How Geologists Tell Time." *Oregon Geology* (November 1991).

Wagener, H. D., and David E. Howell. *Granitic Plutons in the Central and Eastern Piedmont of South Carolina.* Pageland, S.C.: Carolina Geological Society, 1973.

Weekes, Alan F., and John R. Carpenter. *Field Trip Guide to Congaree National Monument.* Columbia: University of South Carolina Center for Science Education, 1993.

Wentworth, Carl, and Marcia Mergner-Keefer. "Regenerative Faults of Small Cenozoic Offset—Probable Earthquake Sources in the Southeastern United States." *Studies Related to the Charleston, South Carolina, Earthquake of 1886—Tectonics and Seismicity.* United States Geological Survey Professional Paper 1313. Reston, Va.: USGS, 1983.

Whitney, James A., et al. "Volcanic Evolution of the Southern Slate Belt of Georgia and South Carolina." *Journal of Geology* 86 (1978).

Willoughby, Ralph. South Carolina Geological Survey. Personal communication, 1993.

Wilson, J. Tuzo. "Continental Drift." *Scientific American* (April 1963).

Windley, Brian F. *The Evolving Continents.* New York: John Wiley and Sons, 1989.

Zupan, Al-Jon. South Carolina Geological Survey. Personal communication, 1993.

INDEX